HARNESSING THE FOURTH INDUSTRIAL REVOLUTION THROUGH SKILLS DEVELOPMENT IN HIGH-GROWTH INDUSTRIES IN CENTRAL AND WEST ASIA

AZERBAIJAN

MAY 2023

ADB

ASIAN DEVELOPMENT BANK

Notes:
In this publication, "$" refers to United States dollars.
ADB recognizes "Hong Kong" as Hong Kong, China; and "Korea" and "South Korea" as the Republic of Korea.

Cover design by Cleone Baradas.

Printed on recycled paper

Contents

Tables, Figures, and Boxes

Boxes

Foreword

The spectacular spread of Fourth Industrial Revolution (4IR) technologies globally has brought great upheavals and disruptions in labor markets. The *2020 Future of Jobs* report by the World Economic Forum estimated that by 2025, 85 million jobs may be displaced by a shift in the division of labor between humans and machines, while 97 million new roles could emerge that address the new realities of division of labor between humans, machines, and algorithms. The erstwhile fears of major job losses have given way to a more balanced discourse on amplifying the promise of technologies for sustainable development and human well-being and minimizing the peril of unemployment. While the new generation of disruptive technologies each have their own unique strength, it is the collective potential of these technologies to improve productivity and the quality of goods and services that have the greatest promise of influencing societal value and impact. Fusing the boundaries between the physical, digital, and biological worlds, the 4IR technologies that include artificial intelligence, robotics, the Internet of Things, 3D printing, genetic engineering, quantum computing, and machine learning, are fast becoming indispensable to modern work life, and indeed to the daily lives of citizens. The question is no longer how to prepare for 4IR technologies tomorrow but how to help individuals, firms, and societies today to effectively draw on them for greater productivity and prosperity.

To provide insights on the impact of 4IR on skills and jobs, the Asian Development Bank undertook the study "Harnessing the Fourth Industrial Revolution Through Skills Development in High-Growth Industries in Central and West Asia." The study suggests concrete pathways by which developing member countries can shape the transition of their economies to 4IR technologies to tap the potential for productivity and new jobs. It provides insights on opportunities, challenges, and promising approaches in 4IR for Azerbaijan, Pakistan, and Uzbekistan with specific focus on two industries in each country deemed important for growth, employment, and 4IR—transportation and storage and agro processing in Azerbaijan, information technology-business process outsourcing and textiles and garment manufacturing in Pakistan, and construction and textiles and garment manufacturing in Uzbekistan. The Central and West Asia region can benefit tremendously by tapping into 4IR technologies. It is important for the region to effectively manage the transition to 4IR technologies for greater economic diversification, moving up the global value chain and strengthening knowledge-based growth processes.

A key aspect of embracing 4IR technologies is to invest in appropriate skills. Based on a recent study by Amazon and Workplace Intelligence, 78% of Gen Z and millennial employees are concerned they lack the skills required to advance their career, and 58% are afraid that their skills have gone stale since the onset of the pandemic, and as many as 70% feel unprepared for the future of work. Hence the time to act on skills development is now, with ever-increasing demand for skills. The coronavirus disease (COVID-19) pandemic has caused incursions in business processes that have led to the acceleration of digital solutions in the marketplace. With the digital talent gap growing, there is a need for both public and private sector entities to invest in re-skilling and upskilling for new and transforming jobs due to adoption of technologies. The study stresses the importance of on-the-job training for 4IR technologies and the need for governments to embark on deliberate strategies for life-long-learning opportunities.

The study affirms a positive outlook to 4IR creating new opportunities for quality jobs. While many jobs will indeed be lost as a result of automation, new jobs will emerge through the adoption of technologies that will increase worker productivity and competitiveness of nations, thereby leading to greater prosperity. However, tapping such benefits is predicated on increasing investments in skills development and greater efforts by companies to upskill their workforce to perform new and higher order roles in complementarity with machines. The study has resulted in a suite of country reports for Azerbaijan, Pakistan, and Uzbekistan and a synthesis report that captures common elements across the three. The reports provide policy makers with evidence-based solutions for skills and talent development to strengthen the countries' readiness for a transition to 4IR.

The study highlights that while job losses will be real, a well-prepared 4IR strategy with industry transformation road maps that are recommended in the study can convert disruptions to opportunities to pivot the workforce to new and modern occupations. In light of post-COVID-19 realities, digital transformation and technology adoption can make enterprises more agile and responsive to changing market conditions.

We believe that 4IR technologies can not only bring greater economic value to enterprises and individuals, they can also help to strengthen the pathways for sustainable and inclusive development. There is more work to be done to explore and leverage the benefits at the intersection of digitalization and climate resilience and to scale up the deployment of 4IR technologies for equity and increasing opportunities for vulnerable populations. We welcome ideas and partnerships with stakeholders as we pursue this area of research toward concrete implementation and next level of analytical work.

Bruno Carrasco
Director General
Sustainable Development and
Climate Change Department
Asian Development Bank

Yevgeniy Zhukov
Director General
Central and West Asia
Department
Asian Development Bank

Preface and Acknowledgments

The Asian Development Bank (ADB) study "Harnessing the Fourth Industrial Revolution Through Skills Development in High-Growth Industries in Central and West Asia" addresses a crucial topic of great relevance to labor markets and jobs. At the heart of this study is the quest to better understand how disruptive technologies are influencing the nature of jobs and skills. Technologies of the Fourth Industrial Revolution (4IR) are influencing every sector and sphere of economies and societies, whether manufacturing or services. At the same time, business processes such as marketing, storage, transport, logistics, and payment mechanisms are greatly transformed with digital technologies. Business practices have been disrupted and reengineered through frontier technologies such as artificial intelligence, digital twins, robotics, and 3D printing.

We bring this piece of research to inform policy makers and practitioners of the implications of 4IR for future job markets. The study drew on various sources of secondary and primary data. It included surveys of employers and training institutions to assess their readiness for 4IR. The study presents analysis of data from online job portals from each of the countries covered in the study – Azerbaijan, Pakistan, and Uzbekistan—to assess trends in skills demand.

The study used a modeling exercise to estimate both job displacement and gains in select industries in the 3 countries. A review of the policy landscape based on benchmarks from international experiences provides the basis for the action points that developing countries can use to harness the potential of Industry 4.0 to increase productivity, facilitate skills development, and incentivize industry. The findings and recommendations from the study underscore the need for renewing skills development strategies with a full life cycle approach. This means that there are no degrees or certificates for life and regular upskilling is essential. The preponderant focus on institution-based training needs to give way to more flexible and multimodal training to include bootcamps, e-learning, and work-place based training. Training for digital skills at basic, intermediate, and higher levels needs a significant ramp up as workplaces undergo digital transformation. The benefits of 4IR can only be effectively harnessed if adequate investments are made in skills development.

The study was led by Shanti Jagannathan, in partnership with Eisuke Tajima and ADB team members. Rie Hiraoka and Brajesh Panth provided valuable guidance to the study. We thank the consultant team led by Fraser Thompson, director, AlphaBeta, for an excellent partnership in this study, together with Wan Ling Koh and Shivin Kohli. AlphaBeta's team developed the analytical model for the study and collaborated closely with ADB's team to bring new insights and directions and we are grateful for this professional collaboration. We thank Xin Long, Aziz Haydarov and Kevin Corbin from ADB headquarters and representatives of ADB resident missions in Azerbaijan, Pakistan, and Uzbekistan, respectively, for their valuable support and country-level consultations (Sabina Jafarova, Sanan Shabanov, Yuliya Hagverdiyeva and Elvin Imanov from Azerbaijan; Khuram Imtiaz and Rizwan Haider from Pakistan; and Farida Djumabaeva and Shahina Rismetova from Uzbekistan). Joehanne Kristal Santos and Evangelyn Medina from ADB provided timely coordination of meetings and activities during the study. Cherry Zafaralla copy edited this report. Dorothy Geronimo coordinated the editorial and publication process with ADB consultants: Maria Theresa Mercado (proofreading), Mariel Gabriel (proof checking), and Edith Creus (typesetting), and Cleone Baradas (cover design).

The study benefited greatly from enriching discussions with government representatives in the respective countries. Early workshops with government representatives and experts were held to inform the study process. The findings of the study were shared in country level workshops. Senior officials and key counterparts consulted are listed at the end of each country report. Tamerlan Tagiyev, Head, Center for Analysis and Coordination of the Fourth Industrial Revolution (Azerbaijan); Shabnum Sarfraz, member, Social Sector and Devolution, Planning Commission, Ministry of Planning, Development and Special Initiatives, Asadullah Faiz, member, Private Sector Development, Punjab Planning and Development Board, and Salman Shami, member, Private Sector Development, and Muhammad Haroon Naseer, additional director general, Punjab Skill Development Authority (Pakistan); and Oybek Shagazatov, head, Main Department of Cooperation with International Financial Organisations, Ministry of Investments and Foreign Trade (Uzbekistan). Several experts also contributed to the study—Amin Charkazov, Ramil Azmammadov (agro-processing) and Sabuhi Abdurahmanov (transport) from Azerbaijan; Allah Bakhsh Malik, Nasir Amin, Muhammad Asim Rehmat (information technology-business process outsourcing) and Muhammad Babar Ramzan (textiles) from Pakistan; and Shukhrathoja Amanov, Khabibullaev Shavkat Azamatovich (construction), and Umida Vakhidova (textile) from Uzbekistan.

We look forward to discussions in taking forward the study's policy recommendations.

Sungsup Ra
Chief Sector Officer
Sustainable Development and
Climate Change Department
Asian Development Bank

Abbreviations

4IR	Fourth Industrial Revolution (or Industry 4.0)
ADB	Asian Development Bank
AR/VR	augmented reality and/or virtual reality
BAU	business-as-usual
COVID-19	coronavirus disease
ICT	information and communication technology
ILO	International Labour Organization
IT–BPO	information technology–business process outsourcing
IOT	Internet of Things
ITM	industry transformation map
SMEs	small and medium-sized enterprises
STEM	science, technology, engineering, and mathematics
TVET	technical and vocational education and training

Executive Summary

The coronavirus disease (COVID-19) pandemic is accelerating the digital transformation of businesses and jobs across all industries. The Asian Development Bank (ADB) *Asian Economic Integration Report 2021* found that accelerated digital transformation can potentially boost global output, trade and commerce, and employment, with Asia expected to reap an economic dividend of more than $1.7 trillion yearly (equivalent to 6.1% of the 2020 regional gross domestic product baseline), or more than $8.6 trillion over the 5-year projection of the study to 2025.

Against this climate, the influence of disruptive technologies on jobs and labor markets has intensified worries around extensive job losses arising from automation and potential disappearance of the comparative advantage of countries based on competitive labor costs. The readiness of developing countries to effectively address the transition to the Fourth Industrial Revolution (4IR) or Industry 4.0 has become an important policy concern. To better understand the implications of 4IR on the future of jobs and to assess the readiness of education and training institutions to prepare workers for future jobs, ADB undertook this study that seeks to capture the anticipated transformations on jobs, tasks, and skills; and to outline policy directions to prepare the workforce for future jobs, particularly in the post-COVID-19 world.

Scope and Methodology of the Study

The study comprises four volumes or reports covering three countries in the Central and West Asia region, namely, Azerbaijan, Pakistan (with a focus on Punjab), and Uzbekistan; including a synthesis report that draws together the findings from the three country studies. This report on Azerbaijan is the first of the four volumes, while the report on Pakistan is volume 2, and volume 3 is the report on Uzbekistan. The synthesis report outlines common areas of policy and action for Industry 4.0.

The study has the following features:

(i) Two focus industries were selected in each country that are crucial for growth, employment, and 4IR: agro-processing and transportation and storage in Azerbaijan; textile and garment manufacturing and information technology–business process outsourcing (IT–BPO) in Pakistan; and textile and garment manufacturing and construction in Uzbekistan.

(ii) For the focus industries, a survey of employers in focus industries, a survey of training institutions, and analysis of data from online job portals to assess trends in skills demand and supply were conducted. The data collected is used to estimate job displacement and gains in the selected industries in each country through a logical model based on economic principles governing job creation and displacement.

(iii)   The policy landscape is also assessed. To understand gaps in the policy landscape in harnessing the potential of 4IR to increase productivity and create quality jobs, this study considered the comprehensiveness of policies in terms of stimulating 4IR adoption and worker reskilling efforts, creating new flexible qualification pathways, and building inclusiveness to extend the benefits of 4IR to all workers. The strength of implementation of policies, particularly against the backdrop of the COVID-19 pandemic, was also assessed.

(iv)   Recommendations on how policy approaches toward 4IR could be strengthened are made. These focused particularly on the investments needed for skills and training, new approaches to deliver training, and other strategies and actions to enhance the readiness of each country's workforce for 4IR going forward.

Surveys of employers and stakeholders were conducted between June and September 2021, and country data from 2020 (the latest for which full-year data is available) were used. To align all baselines for comparison, the survey data collected is assumed to reflect perspectives and circumstances as of end of 2020. The objective is to provide an illustrative view of how 4IR can impact jobs and skills in the three countries in the Central and West Asia region between 2020 and 2025.

Key Findings for Azerbaijan

This report covers the key findings for Azerbaijan. The report analyzes the implications of 4IR for jobs, tasks, and skills in the agro-processing industry, as well as the transportation and storage industry. These industries were selected based on several factors including their importance to national employment and growth prospects, the degree of relevance of 4IR technologies to the industry, and alignment with regional and national growth plans. The development of the agriculture and agro-processing industries forms a key part of Azerbaijan's plans to build its non-oil economic sectors and diversify its economy. The agro-processing sector employed only 1.4% of the national workforce in 2018 but is one of the country's fastest growing industries in terms of employment.

Industry 4.0 technologies present significant opportunities for productivity gains across the agro-processing value chain. The *Strategic Vision and Roadmap for Azerbaijan Agriculture* published in 2016 sets out plans to transform Azerbaijan's agro-processing industry through the application of modern technologies. Similarly, Azerbaijan's transportation and storage industry has significant growth potential. The road map thus also sets out plans to transform Azerbaijan into a regional trade and logistics hub by strengthening infrastructure and identifying performance bottlenecks, as well as attracting foreign investments. In 2018, the transportation and storage industry contributed 4.2% to national employment and there is potential for this to increase further. 4IR technologies can play a role in the transformation of the transport and logistics value chain, with past research in Asia suggesting an increase in labor productivity of up to 1.14% for every 1% increase in uptake of artificial intelligence (AI) technologies.

The report finds that 4IR will have a transformational effect on jobs and skills in the agro-processing, and transportation and storage industries in Azerbaijan with strong potential for positive gains in jobs and productivity, which can be reaped through adequate investments in skills and training. Key findings from the country report include the following:

(i)   **Firms in both industries require support to adopt 4IR technologies in their operations.**

     (a)   Only 14% of agro-processing firms and 20% of transportation and storage firms surveyed have a good understanding of 4IR technologies and their applications, although many are keen to adopt technologies such as the Internet of Things (IOT) to a larger extent by 2025.

(b)   There are strong expectations regarding potential gains from adopting 4IR technologies. Agro-processing firms estimate that the adoption of 4IR technologies will increase labor productivity (i.e., output per worker) by 49% in between 2020 and 2025 while transportation and storage firms estimate a 41% increase.

(ii)   **The full adoption of 4IR technologies can bring net job gains in the agro-processing and transportation and storage industries in Azerbaijan.**

(a)   The report estimates that the adoption of 4IR technologies in the agro-processing and transportation and storage industries will create net job gains by comparing the displacement and productivity effects of adopting 4IR technologies. The number of new jobs created by productivity gains from adopting 4IR technologies will exceed the number of jobs displaced by automation.

(b)   As a result of 4IR technologies adoption, more jobs are expected to be created—close to 15,000 new jobs in Azerbaijan's agro-processing industry and around 27,000 new jobs created in the transportation and storage industry. This is over and beyond the business-as-usual job growth in a scenario without 4IR adoption, meaning that these jobs are fully attributable to 4IR adoption. These jobs gains can only be realized if policy makers adopt policies to support the adoption of 4IR technologies in firms as well as build a 4IR-ready workforce able to support the industry's transformation.

(iii)   **Manual jobs will be displaced, and technical jobs will be created.**

(a)   The new jobs created by 4IR will not be identical to the jobs displaced with employers in both industries. The assessment shows that the number of technical roles is likely to see the largest increase with 4IR adoption and the largest displacement is expected in manual roles. This implies that workers displaced by 4IR technologies are likely to be from manual roles that often require lower levels of education and skills, and pay less well, while jobs created are likely to be in better paid technical roles that require higher levels of education.

(b)   The change in the types of jobs available will have several implications for Azerbaijani workers and firms. On one hand, it means that better jobs will be available to Azerbaijani workers across the agro-processing and transportation and storage industries. On the other hand, it means that one, strong social protection and reskilling policies would be needed to ensure that manual workers are not adversely impacted by the adoption of 4IR technologies; and two, without a skilled workforce to support the adoption of 4IR technologies, firms cannot complete the transition toward Industry 4.0.

(c)   Gains from 4IR will not be equitably distributed among male and female workers. The number of new jobs expected to be gained by male workers is 1.6 times that expected to be gained by female workers for the agro-processing industry and 1.8 times for the transportation and storage industry. Efforts must therefore be taken to ensure that female workers are not disadvantaged by increased automation and that they also benefit from the gains of Industry 4.0.

(iv)   **Alignment on industry's skills needs between employers and training institutions is weak in Azerbaijan** .

(a)   The adoption of 4IR technologies will change the skill needs of employers in both industries by 2025. For agro-processing and transportation and storage firms, information and communication technology (ICT) and digital skills, as well as creative thinking and/or design skills, will be increasingly valued. Training institutions should be equipped to prepare workers for these skills changes.

(b) There are gaps in graduate quality perceived by training institutions and employers. While 83% of training institutions assess their graduates to be adequately prepared for entry-level positions, less than a quarter of employers in the agro-processing and transportation and storage industries take the same view. Most employers view graduates in Azerbaijan to be lacking both general skills (e.g., soft skills, creativity) as well as job-specific skills.

(c) Collaboration in the design and implementation of training curricula and programs is weak between training institutions and industry stakeholders in Azerbaijan. Currently, only 20% of training institutions gather input from industry stakeholders to design curricula, and only 44% work with employers to provide industry placements for staff for training purposes. While over half of employers surveyed in both industries plan to adopt IOT technologies, only 4% of training institutions offer courses relevant to IOT. Industries should be more involved in the training of both students and staff to ensure that teaching staff have the relevant knowledge to produce 4IR-ready students.

(v) **Training institutions require additional financial and technical support to be 4IR-ready.**

(a) About 54% of training institutions surveyed strongly agree that technical and financial support is needed to enable them to prepare workers for 4IR. As of 2020, only 13% of training institutions surveyed currently teach courses related to 4IR and only 29% provide general digital skills programs to improve digital literacy. Less than half of training institutions surveyed organize ongoing professional development and training for instructors. Training institutions require technical and financial support to build up the capabilities of their teaching staff to ensure that they are equipped to teach 4IR-related courses as well as implement 4IR-enabled teaching approaches in the classroom to ensure that graduates are familiar with the use of advanced technologies.

(vi) **Government, industry, and training institutions can further strengthen their coordination to ensure that Azerbaijan can reap the gains of 4IR.**

(a) Azerbaijan has made significant efforts to update and modernize its curricula in recent years and build a framework for continual reskilling. Sectoral skills councils that include employers, policy makers, and training institutions have been established and the National Qualifications Framework for Lifelong Learning adopted. The Government of Azerbaijan recognizes the potential of 4IR technologies in promoting growth and a Center for Analysis and Coordination of the Fourth Industrial Revolution was established in early 2021.

(b) Despite these efforts, the quality and depth of coordination will need to be strengthened among training institutions, employers, and government. The training institution and employer surveys highlighted several mismatches in terms of views on quality of graduates, engagement in curriculum design (particularly around specific courses desired by employers but not offered by training institutions), and level of industry knowledge of teaching staff. Many employers also lack an understanding of 4IR technologies and are not preparing their workers accordingly. There is also a need to increase incentives for employers to engage in training, as well as introduce programs targeted at new types of workers such as digital freelancers and gig economy workers.

Key Recommendations and Way Forward

Drawing on the findings of industry and training institution surveys conducted as well as the policy assessment, this country report identifies eight recommendations for Azerbaijan to strengthen its preparedness toward 4IR. These include recommendations to strengthen both the demand for skills (i.e., creation of 4IR-related jobs) and the supply of skills. The specific actions for each of the eight Azerbaijan recommendations include the following:

(i) **Develop 4IR adoption plans to complement existing sectoral road maps.** To better integrate 4IR trends with industry development plans and skills development needs, Azerbaijan could consider the development of 4IR action plans modelled after Singapore's Industry Transformation Maps (ITMs), which provide information on technology impacts, labor market shifts, skills required for different occupations, and reskilling options for different industries. These action plans could build on the existing strategic road maps to focus on how the adoption of 4IR technologies can help to achieve long-term growth objectives. The plans should identify the impact of adopting 4IR technologies on the current workforce, and methods to strengthen the workforce to enable a smooth transition toward 4IR. The Center for Analysis and Coordination of the Fourth Industrial Revolution could take the lead in coordinating these plans, which could be implemented in 12–36 months.

(ii) **Develop programs to build awareness of digital tools among firms.** The employer surveys revealed that only 14% of agro-processing firms and 20% of transportation and storage firms have a good understanding of 4IR technologies as of 2020. In Germany, the Mittelstand 4.0 Competence Centers have been established with the support of the Federal Ministry of Economics and Technology to help businesses gauge their current stage of digitalization, and jointly develop and implement digitalization road maps and solutions. Similar efforts in Azerbaijan could be led by the Center for Analysis and Coordination of the Fourth Industrial Revolution, building on the Center's ongoing efforts to help firms adopt digital technologies. This could potentially be implemented in less than 12 months.

(iii) **Implement incentive schemes for firms to train employees for 4IR.** Incentive schemes to encourage firms to provide training to workers could contribute to building a 4IR-ready workforce as well as reduce the risk of some groups of workers being displaced by automation. In particular, incentives could be focused on small and medium-sized enterprises (SMEs) that face the most significant resource constraints in sending workers for training, particularly as over half of all enterprises in Azerbaijan are small enterprises with fewer than 25 employees. In Malaysia, a Skills Upgrading Program provides grants covering 70% of training fees for technical and soft skills for SMEs to train their workers. The Small and Medium Business Development Agency of Azerbaijan could lead efforts to identify the specific challenges faced by SMEs in providing training to their workers and design programs to counter these challenges. This could be implemented in 12–36 months.

(iv) **Provide programs to strengthen industry knowledge and digital skills of trainers and teachers.** Less than half of training institutions in Azerbaijan currently provide professional development and training for instructors to update their knowledge of the latest equipment used by industry and skills required. To ensure that training institutions are adequately prepared for 4IR, the Ministry of Education could take the lead in programs to improve the competencies of teaching staff in schools and training institutions at various levels. For instance, in Malaysia, INTI International University & Colleges created the Faculty Industry Attachment (FIA) program, which enables teaching staff to work with industries as part of their regular working hours to broaden their practical experiences and stay abreast of the latest developments in the industry. Similar industry attachment programs for teaching staff of vocational training institutions in Azerbaijan could improve industry knowledge. Programs to build strong digital literacy among teaching staff could also be implemented at all levels, to ensure that students in Azerbaijan form strong fundamentals in digital literacy from an early age. This could potentially be implemented in 12–36 months.

(v) **Develop online learning platforms.** The adoption of 4IR technologies will rapidly and constantly change the skills and practical knowledge required of workers. In Azerbaijan, there are limited opportunities for adult learning due to the lack of adequate adult training facilities and courses, particularly outside of urban centers. Azerbaijan could consider online-learning platforms to rapidly build up the new skills required by employers. Under the Republic of Korea's Life-Long Learning Promotion Plan (2018–2022), online learning platforms have been established to upskill the population in a range of areas, including digital skills. One example is the Korean massive open online courses or MOOCs. Since its launch in 2015, over 1,700 accredited courses at the higher education level have been developed through partnerships with local universities, with a significant share of them focused on advanced digital courses such as machine learning, AI navigation and perception, and mathematics for data scientists. In Azerbaijan, online learning platforms could similarly be used to upskill the workforce. The use of online platforms would allow for cheaper course offerings and a wider reach for training courses. The Ministry of Education of Azerbaijan could take the lead in developing suitable platforms, working with the State Agency for Public Services and Social Innovations and other relevant agencies. This could potentially be implemented in less than 12 months.

(vi) **Develop innovative job-matching initiatives and platforms.** Interviews and consultations with local experts and stakeholders revealed that the lack of information on available jobs and skills required by employers is a key barrier to workers securing quality jobs in Azerbaijan. Only 4% of training institutions in Azerbaijan work with employers to organize job fairs to advertise job opportunities. Policy makers can consider innovative approaches to improve job-matching between employers and prospective workers. The Ministry of Labor and Social Protection of the Population could consider working with technology agencies such as the State Agency for Public Services and Social Innovations to launch platforms incorporating AI or Big Data technologies, which could help job seekers or recruiters sieve through the opportunities or candidates available easily. This could potentially be implemented in less than 12 months.

(vii) **Develop skilling and labor support programs for digital freelancers.** One key aspect of the digital economy and 4IR is the proliferation of digital freelancers. There is significant potential for young Azerbaijanis to find quality jobs as digital freelancers with the rise of the global freelance economy against the backdrop of the COVID-19 pandemic. Various initiatives could be pursued in Azerbaijan to strengthen the digital freelancing capabilities of young adults and enable them to find quality freelance jobs. In particular, the Innovation Agency of Azerbaijan (together with the Ministry of Transport, Communication and High Technologies) could lead efforts to provide training to youth in areas such as digital marketing and website development, as well as create an enabling environment for digital freelancers through fiscal incentives and micro-certification programs. This could potentially be implemented in 12–36 months.

(viii) **Develop programs to enable more women to take up technical jobs.** This country report estimates that the number of new jobs from the adoption of 4IR technologies expected to be gained by male workers is 1.6 times that expected to be gained by female workers for the agro-processing industry, and 1.8 times for the transportation and storage industry. As the adoption of 4IR technologies creates new jobs largely in technical roles, targeted programs would need to be developed to enable and encourage more female workers to take up technical jobs. In Indonesia, the Philippines, and Thailand, the International Labour Organization (ILO)'s Women in Science, Technology, Engineering, and Mathematics (STEM) Workforce Readiness and Development Program aims to enhance the employability of women for STEM-related jobs (ILO 2020a). In Australia, the Curious Minds program funded by the Department of Education, Skills and Employment is aimed at highly capable female students (aged around 15–16 years old) who have an interest in STEM. Policy makers in Azerbaijan could consider similar initiatives, in collaboration with international partners, to strengthen the capabilities and interest of women in taking on up technical professions so that they will benefit from the job gains from 4IR. The Center for Analysis and Coordination of the Fourth Industrial Revolution could lead efforts to ensure that female workers can also benefit from 4IR. This could potentially be implemented in 12–36 months.

While these recommendations apply to both the agro-processing and transportation and storage industries, a set of priorities unique to each industry should be considered when implementing the respective recommendations.

Agro-Processing Industry

Only 14% of agro-processing firms in Azerbaijan reported a good understanding of 4IR technologies as of 2020 and less than 20% of firms expect to adopt technologies such as autonomous robots, additive manufacturing, and Big Data analytics across various functions by 2025. As such, to reap the gains of 4IR, policy makers would need to play an active role in devising 4IR adoption plans that complement existing sectoral road maps and support companies to deploy 4IR technologies. This report shows that in making the transition toward 4IR, the lack of quality training providers and impact on women will be particularly challenging for Azerbaijan's agro-processing industry. Over 60% of employers surveyed disagreed that it is easy to find good quality trainers, and the number of new jobs expected to be gained by male workers is 1.6 times that expected to be gained by female workers. Targeted programs would need to be adopted to ensure that trainers and teachers in the country are able to support the agro-processing workforce to meet the new skill needs created by 4IR. As roles in the agro-processing industry shift toward more technical occupations dominated by men, active intervention would also be needed to ensure that female workers can benefit from 4IR adoption and programs developed for women to take up technical jobs.

Transportation and Storage Industry

Like the agro-processing industry, transportation and storage companies in Azerbaijan also have a limited understanding of 4IR with only 20% of firms surveyed reporting a good understanding of 4IR and their applications. However, a larger proportion of firms expect to adopt various 4IR technologies such as systems integration, blockchain technology, and IOT technology by 2025. As such, it would be integral for policy makers to support firms in making the transition toward 4IR by developing programs and 4IR competency centers to build awareness of digital tools. In tandem with the shift toward 4IR technology adoption, the job scope of workers in the transportation and storage industry will change as would skill needs. Most jobs created will be in technical roles in which the workforce might not be equipped for, and significant changes in skill demand may lead to challenges in hiring workers. In particular, digital or ICT skills, creative thinking and/or design, and numeracy skills will become more important with 4IR adoption. Policy makers could support reskilling of workers through incentive schemes for firms to train employees and through development of online learning platforms.

1 The Industry 4.0 Skills Challenge

This chapter investigates the demand and supply of skills driven by the adoption of Fourth Industrial Revolution (4IR or Industry 4.0) technologies such as Internet of Things, artificial intelligence, additive manufacturing, robotics, and Big Data for both the agro-processing and transportation and storage industries in Azerbaijan. The analysis uses a range of data, including employer surveys, expert interviews, online job board data, and national labor market statistics, and projects implications from 2020 to 2025.

The analysis shows the significant potential of 4IR technologies to create quality jobs for workers in Azerbaijan's agro-processing and transportation and storage industries if robust policies are implemented to educate firms on the benefits of 4IR and build a 4IR-ready workforce. However, it also cautions that the impact of job gains and losses are not evenly distributed across occupational groups, and that workers engaged in manual, basic jobs are most likely to be adversely impacted. As such, it would be important to build strong social protection policies as well as reskilling programs, supported by robust implementation mechanisms to support these workers.

The report points to the potential of 4IR to create net job gains in both industries. Close to 15,000 more jobs or the equivalent of 20% of the 2020 agro-processing industry workforce, over and beyond business-as-usual growth rates, are expected to be created in the agro-processing industry by 2025. Over 27,000 more jobs (13% of the 2020 industry workforce) are expected to be created in the transportation and storage industry by 2025, if firms fully adopt 4IR technologies by then. However, 20% or less of firms in each industry have a good understanding of 4IR technologies and their applications, so these jobs gains cannot be realized without significant intervention by the government.

Alongside the creation of 4IR-related jobs, the report also finds that the skills required in these jobs are different from the skills that employers seek in workers in 2020. Digital and/or information and communication technology skills, with creative thinking and/or design skills, will be increasingly valued by employers by 2025. Only a small proportion of firms in the agro-processing and transportation and storage industry continuously train workers, and programs to incentivize and provide them with the necessary resources to do so could be needed. As such, strong reskilling policies would be needed to update workers' skills and ensure that they are 4IR-ready.

A.   Industry 4.0 and Its Relevance for Azerbaijan

The Fourth Industrial Revolution (4IR) or Industry 4.0 is poised to fundamentally change the future of work. 4IR can be described as the advent of "cyber-physical systems" involving entirely new capabilities for people and machines,[1] wherein new technologies, such as the Internet of Things (IOT), artificial intelligence (AI), additive manufacturing,

---

[1] World Economic Forum. What is the Fourth Industrial Revolution? https://www.weforum.org/agenda/2016/01/what-is-the-fourth-industrial-revolution/.

robotics, and Big Data analytics among others, become embedded within societies. 4IR is fundamentally different from past industrial revolutions in its potential implications for economies and the workforce.

What could 4IR mean for Azerbaijan? The Government of Azerbaijan recognizes the potential of 4IR and the need to strengthen the country's human capital to enable technology adoption. In 2016, Azerbaijan launched a series of strategic road maps that set out the government's plans for inclusive economic development for 2025 and beyond, focusing on accelerating growth in priority industries, such as agriculture and the transportation and storage industries, through technology adoption and skills development (Government of the Republic of Azerbaijan 2016f). The Center for Analysis and Coordination of the Fourth Industrial Revolution under the Ministry of Economy—a new public entity tasked to coordinate Azerbaijan's response to 4IR trends—was established in early 2021 (Azerbaijan Press Agency 2021).

Understanding how the skills landscape is likely to change under 4IR is becoming more difficult with the rapid pace at which technology is developing and being adopted. This is particularly so as these changes have accelerated against the backdrop of the coronavirus disease (COVID-19) pandemic. This means traditional approaches of assessing skill gaps, often relying on time-intensive processes to collect data that quickly become outdated, may no longer be suitable. This report explores a new approach to understanding the labor market implications of 4IR. Some of the key design aspects examined are as follows:

(i)   **Use of primary and secondary local data.** This report utilizes a variety of local data sources, including data from the State Statistical Committee of the Republic of Azerbaijan, surveys of employers and training institutions in the agro-processing and transportation and storage industries in Azerbaijan, and interviews with local experts and key stakeholders. A summary of the primary data sources used is in Table 1.

Table 1: Primary Data Sources Used

Employer surveys	A survey of 50 transportation and storage firms and 50 agro-processing firms in Azerbaijan was undertaken in Azerbaijan. The surveys were carried out by M-Vector, a market research company.
Training institution surveys	A survey of 70 training institutions was undertaken in Azerbaijan. These included public and private institutions of higher learning as well as technical and vocational education and training institutions. Of the training institutions surveyed, 97% trained at least 100 students per year. The surveys were carried out by M-Vector.
Online job portal analysis	The analysis covered 30 job listings in the agro-processing industry and 77 job listings in the transportation and storage industry obtained from the job portal Offer.AZ (accessed June 2021).

Sources: Asian Development Bank (Sustainable Development and Climate Change Department) and AlphaBeta.

(i)   **Use of current market information.** Given the rapid changes in the labor market, labor market surveys can become quickly obsolete. To provide an updated snapshot of skills needs, this report uses information on skill profiles for current jobs advertised in major online job portals.[2] Machine learning algorithms were applied to analyze data obtained from local job portals to understand the skills demanded by employers in the two industries.

(ii)   **Focus on both demand and supply.** The report considers changes in the demand of skills brought about by the adoption of 4IR technologies, as well as the supply of skills, and the readiness of the training landscape to upskill and reskill workers for 4IR.

---

[2]   The analysis covered 30 job listings in the agro-processing industry and 77 job listings in the transportation and storage industry from the job portal Offer.AZ (accessed June 2021).

The surveys of employers and stakeholders were conducted between June and September 2021, and country-level data from 2020 (the latest for which full-year data is available) was used. To align all baselines for comparison, the survey data collected is assumed to be reflective of perspectives and circumstances as of end of 2020. The objective is to provide an illustrative view of how 4IR can impact jobs and skills in the three countries in the Central and West Asia region between 2020 and 2025.

B.   Industry Selection

Two industries were selected to conduct further analysis of the implications of 4IR adoption for the demand and supply of skills. A two-step methodology was used to select the industries:

(i)   **Shortlisting industries prioritized by the Government of Azerbaijan for future growth or for 4IR application.** This included reviewing the Strategic Roadmap for the National Economy Perspective of the Republic of Azerbaijan and Strategic Roadmap for the Development of Telecommunications and Information Technologies in the Republic of Azerbaijan, among other policy documents.

(ii) **Scoring and ranking.** Shortlisted industries were scored and ranked according to a set of criteria:

(a)   How significant is the industry's contribution to the country's employment?

(b)   Does it exhibit strong recent employment growth?

(c)   Are its exports internationally competitive?

(d)   How relevant are 4IR technologies to the industry?

(e)   Is the relevant data available for the industry analysis?

The shortlisted industries were then tested with stakeholders across government, industry, and civil society during a consultation workshop conducted in May 2021. Based on this process, the following industries were selected for the analysis:

(i)   **Agro-processing.** The development of the agriculture and agro-processing industries forms a key part of Azerbaijan's plans to build its non-oil economic sectors and diversify its economy. In 2018, agro-processing employed only 1.4% of the national workforce but is one of the country's fastest growing industries in terms of employment.[3] Employment in the industry grew at 4.8% on average between 2013 and 2018, outperforming the national employment growth rate of 1.5%, and only lagging the tourism industry (footnote 3). The *Strategic Vision and Roadmap for Azerbaijan Agriculture* published in 2016 highlights plans to transform the agro-processing industry through the application of modern technologies, to capture the higher value of processed food products as compared to raw materials (Government of the Republic of Azerbaijan 2016a). The road map sets out a long-term vision of strengthening agro-processing value chains and building a competitive advantage in the production and export of value-added industrial crops, including cotton, herbal tea, and tobacco, by 2025. Industry 4.0 technologies such as IOT, AI, additive manufacturing, autonomous robots, and Big Data analytics present significant opportunities for productivity gains across the agro-processing value chain, and it is timely for policy makers in Azerbaijan to consider how these technologies can be used to achieve Azerbaijan's economic diversification and growth objectives.

---

[3]   Sources: Statistical Committee of Azerbaijan, ILO statistics, WTT council, AlphaBeta analysis.

(ii) **Transportation and storage.** Azerbaijan is located along key trade corridors between Asia and Europe and has significant potential to capture transit trade as well as boost import and export volumes. In 2018, the transportation and storage industry contributed 4.2% to national employment and this has potential to increase further. The *Strategic Roadmap for the Development of Logistics and Trade in the Republic of Azerbaijan* released in 2016 sets out plans to transform Azerbaijan into a regional trade and logistics hub by strengthening infrastructure and identifying performance bottlenecks, as well as attracting foreign investments (Government of the Republic of Azerbaijan 2016b). In 2020, the Port of Baku handled over 4.8 million tons of cargo, a 20% increase compared to 2019, with transit cargo taking up 87% of total cargo (Port of Baku 2021). 4IR technologies can play a role in the transformation of the transport and logistics value chain with past research in Asia suggesting an increase in labor productivity of up to 1.14% for every 1% increase in uptake of AI technologies (PWC 2018). It is timely for policy makers to consider how 4IR technologies can help to achieve Azerbaijan's goals to become a regional trade and logistic hub.

C.   Agro-Processing

Past research estimates that the adoption of 4IR technologies could bring productivity gains of around 50% in the food industry (McKinsey & Company 2018). In Azerbaijan, despite firms having a limited understanding of 4IR technologies (only 14% of firms report an intermediate or advanced understanding of 4IR technologies and their applications), the agro-processing firms have positive expectations of the impact of technology and expect labor productivity (as measured by output per worker) to increase by 49% on average by 2025 with the adoption of 4IR technologies.

If these productivity gains are realized across the entire agro-processing industry in Azerbaijan, a significant number of new jobs will be created by 2025. This report estimates that around 15,000 net new jobs or the equivalent of 20% of the 2020 agro-processing workforce could be created by 2025, over and beyond the number of jobs created by business-as-usual (BAU) growth. The 15,000 net new jobs created are a function of 45,000 jobs created and 30,000 jobs displaced with the adoption of 4IR technologies. In other words, this does not mean that no workers will be displaced by automation but that the number of jobs created by the overall growth of the industry due to 4IR adoption will exceed those displaced. However, this report also shows that those who are likely to lose their jobs (in more manual roles) may not seamlessly transition into new jobs created (which are generally more technical in nature) without significant reskilling. There is also a concerning lack of worker training being provided by firms as of 2020, creating risks where the benefits of 4IR technologies may not be reaped, and the job creation potential not realized.

Relevance of Industry 4.0

Previous research identified a range of technologies that could transform every aspect of the agro-processing value chain, from using sensors or data-driven solutions, to reducing machine downtime, to leveraging AI to track consumer preferences (McKinsey & Company 2018).

Some key technologies include the following:

(i) **Internet of Things.** The IOT refers to networks of sensors and actuators embedded in machines and other physical objects that connect with one another and the internet. It has a wide range of applications, including data collection, monitoring, decision making, and process optimization (McKinsey Global Institute 2014). IOT systems and technologies such as sensors and smart shelves can be used to gauge inventory and optimize stock as well as allow better regulation of temperature, humidity, and other factors across the manufacturing process. Candymaker Hershey uses IOT sensors to monitor the temperature of licorice and better predict the net weight of the end product so that it can downsize its products without breaching regulatory guidelines. It estimates that a 1% downward adjustment in size can result in $0.5 million savings per 14 gallon batch of licorice (Maddox 2017).

(ii) **Artificial intelligence.** The ability of machines to learn and act intelligently and carry out a wide range of human-like processes is a product of AI. This means machines can make decisions, carry out tasks, and even predict future outcomes based on what they learn from the data. Among other uses, AI technology can be used to sort feedstock of different grades and maintain cleanliness and food safety compliance, as well as conduct quality checks on food products (Sharma 2019). An AI-based cleaning system developed by the University of Nottingham could save food manufacturers in the United Kingdom (UK) up to £100 million a year. The system optimizes the cleaning process for food manufacturing equipment by monitoring the amount of food and microbial debris on the equipment. This would reduce the energy and water use as well as downtime for manufacturers by 20%–40% (University of Nottingham 2016).

(iii) **Additive manufacturing (three-dimensional printing).** Additive manufacturing technologies produce physical objects from digital models by adding thin layers of material in succession. There are several applications of three-dimensional (3D) printing in food manufacturing—from producing specialized spare parts for food processing machinery, to creating novel food products. Beer company Heineken uses 3D printing to create custom parts and tools for its production line. With 3D printing, a tool loosens and tightens the columns of guiding wheels that apply bottle labels, yielding cost savings of 70% for the company, and reducing delivery time from 3 days to 1, compared to traditional production methods.[4] New food products can also be created using 3D printing. In Singapore and Sweden, researchers are applying 3D printing technique to create meals for elderly and frail patients who have trouble swallowing food (Ng and Enriquez 2020).

(iv) **Autonomous robots.** Autonomous robots are intelligent machines capable of performing tasks with a high degree of autonomy. They can be used to automate various parts of the agro-processing workflow including sorting of raw materials or products, cutting of ingredients, and product packaging. For instance, the fruit packer Adrian Scripps in the UK uses a robotic crate packing solution to pack apples without bruising them, leading to a threefold improvement in productivity per person (South East Farmer 2021).

(v) **Big Data analytics.** Big Data analytics is the use of advanced analytic techniques on very large, diverse data sets. Applications include predictive analytics for maintenance and repair of production lines and analysis of consumer data to predict consumer buying patterns or create new products. For instance, predicative analytics can help to avoid expensive downtime, for example, a single hour of downtime can cost between $100,000–$5 million across industries (McKinsey & Company 2018b). Coca Cola used consumer data from their self-service softdrink fountains, which allowed customers to prepare drinks by mixing a variety of flavors to create and launch the Cherry Sprite (Marr 2017).

A total of 50 agro-processing firms in Azerbaijan were surveyed for the employer surveys. Despite the numerous applications of various 4IR technologies in the agro-processing and food manufacturing industry, the employer surveys show that agro-processing firms in Azerbaijan currently have a limited understanding of 4IR technologies. Only 14% of firms surveyed indicated an intermediate or advanced level of understanding of 4IR technologies and the potential benefits that such technologies could bring to their industry (Figure 1). This contrasts with a previous study in Viet Nam that found that more than half of Vietnamese firms in the agro-processing industry reported a good understanding of 4IR technologies (ADB 2021). Half of firms surveyed also felt that firms in their

---

[4]   AMFG. How 3D Printing Transforms the Food and Beverage Industry. https://amfg.ai/2020/08/17/how-3d-printing-transforms-the-food-and-beverage-industry/.

Figure 1: Employers' Understanding of Industry 4.0 Technologies in the Agro-Processing Industry in Azerbaijan (%)

Only 14% of employers in the agro-processing industry believe that they have a good understanding of 4IR technologies

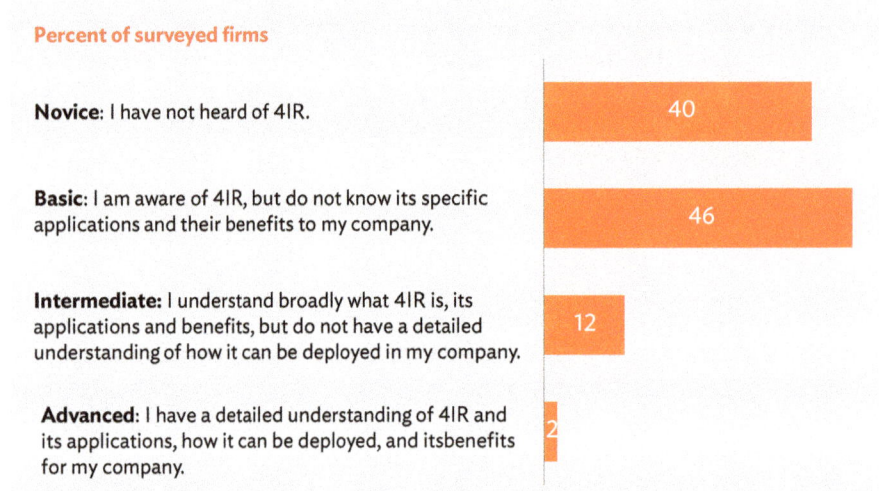

Percent of surveyed firms

Novice: I have not heard of 4IR. — 40

Basic: I am aware of 4IR, but do not know its specific applications and their benefits to my company. — 46

Intermediate: I understand broadly what 4IR is, its applications and benefits, but do not have a detailed understanding of how it can be deployed in my company. — 12

Advanced: I have a detailed understanding of 4IR and its applications, how it can be deployed, and its benefits for my company. — 2

4IR = Fourth Industrial Revolution.
Note: Based on survey of employers in the agro-processing industry between June and September 2021 (n=50).
Source: Asian Development Bank (Sustainable Development and Climate Change Department).

supply chain have a limited understanding of 4IR technologies and did not use such technologies (Figure 2). This suggests that 4IR adoption is relatively limited along the agriculture and agro-processing value chain, including among farmers as well as retailers. Robust policies and programs to improve firms' understanding and adoption of the 4IR technological tools available are needed to ensure that agro-processing firms and workers in Azerbaijan can reap the gains of Industry 4.0.

Despite having a limited understanding of 4IR technologies, agro-processing firms in Azerbaijan appear to have a positive impression of technology and expect 4IR technologies to bring significant productivity gains. Past research estimates that the adoption of 4IR technologies could bring productivity gains of around 50% in the food industry (McKinsey & Company 2018b). In Azerbaijan's agro-processing industry, firms similarly expect labor productivity to increase by 49% on average by 2025, with the adoption of 4IR technologies (Figure 3).

The strong expectations of labor productivity gains could be due to the limited adoption of 4IR technologies as of 2020 and the plans of most firms to at least experiment with these technologies by 2025 (Figure 4). While most firms currently do not adopt various 4IR technologies identified to be critical to the agro-processing industry, over 70% of firms have plans to adopt each technology by 2025, even if on a limited scale within their company.

Firms in the agro-processing industry are particularly interested in IOT technologies. In 2020 (based on survey data from 2021), 40% of firms do not use IOT technologies at all, but 98% of firms plan to use IOT technologies by 2025. This could be due to IOT technologies having numerous applications in the food manufacturing industry—from the use of devices that monitor and regulate temperature and humidity across the manufacturing process, to smart shelves that can be used to keep track of inventory. Agro-processing firms are also keen to

Figure 2: Employers' Understanding of Industry 4.0 Technologies among Firms in the Agro-Processing Supply Chain in Azerbaijan (%)

Half of agro-processing firms feel that companies in their supply chain have limited understanding of 4IR technologies

Percent of surveyed firms

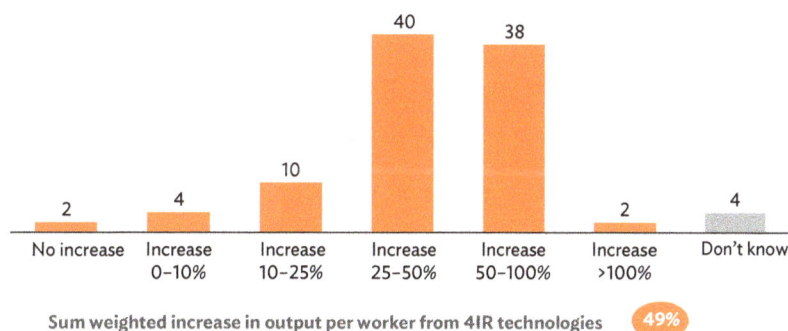

Novice: Our supply chain companies have not heard of 4IR.   32

Basic: Our supply chain companies are aware of 4IR technologies, but do not use them.   18

Intermediate: Our supply chain companies have some understanding of 4IR technologies, and either use them on a small scale or plan to do so in the near future.   22

Advanced: Our supply chain companies have a detailed understanding of 4IR technologies, and usethem in their operations.   0

We do not know   28

4IR = Fourth Industrial Revolution.
Note: Based on survey of employers in the agro-processing industry between June and September 2021 (n=50).
Source: Asian Development Bank (Sustainable Development and Climate Change Department).

Figure 3: Expected Increase in Output Per Worker Due to Industry 4.0 Technologies by 2025 in the Agro-Processing Industry in Azerbaijan (%)

The majority of agro-processing firms expected 4IR technologies to increase output per worker by more than 25% in 5 years' time

Percent of surveyed firms

No increase	Increase 0–10%	Increase 10–25%	Increase 25–50%	Increase 50–100%	Increase >100%	Don't know
2	4	10	40	38	2	4

Sum weighted increase in output per worker from 4IR technologies   49%

4IR = Fourth Industrial Revolution.
Notes: Based on survey of employers in the agro-processing industry between June and September 2021(n=50)· Calculated using sum-weighted average of output increase by the number of firms indicating different levels of expected increase in output, i.e., 0%, 0%–10%, 10%–25%, 25%–50%, 50%–100%, and over 100%. The midpoint of the range for each option for expected increase in output is used; for expected output increase of over 100%, the lower bound of 100% is used.
Source: Asian Development Bank (Sustainable Development and Climate Change Department).

Figure 4: Current and Future Adoption of Relevant Industry 4.0 Technologies in the Agro-Processing Industry in Azerbaijan (%)

Compared to other 4IR technologies, agro-processing firms expect to increase the deployment of Internet of Things technologies most significantly

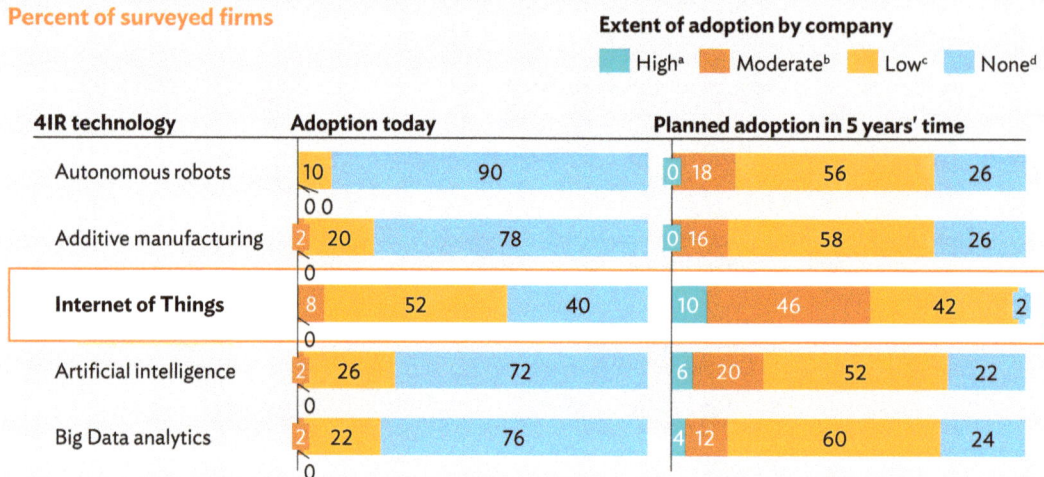

Percent of surveyed firms

Extent of adoption by company: High[a] Moderate[b] Low[c] None[d]

4IR technology	Adoption today				Planned adoption in 5 years' time			
Autonomous robots	10	90	0	0	0	18	56	26
Additive manufacturing	2	20	78	0	0	16	58	26
Internet of Things	8	52	40	0	10	46	42	2
Artificial intelligence	2	26	72	0	6	20	52	22
Big Data analytics	2	22	76	0	4	12	60	24

Notes: Based on survey of employers in the agro-processing industry between June and September 2021 (n=50).
a "High": Firm has fully deployed the technology across all possible functions in the enterprise and/or has plans to fully deploy the technology across all possible functions in the future.
b "Moderate": Firm has implemented the technology, but not fully deployed across all possible functions in the enterprise and/or plans to implement the technology across a few functions in the future.
c "Low": Firm is experimenting with the technology at a very limited scale within the enterprise and/or plans to experiment with the technology in the future.
d "None": Firm has not used technology at all within the enterprise and/or has no plans to use the technology in the future.
Source: Asian Development Bank (Sustainable Development and Climate Change Department).

adopt other types of 4IR technologies, with over a quarter looking to deploy AI technologies across a range of functions. The use of AI technologies can improve not only labor productivity but also increase the quality of products. For example, Japan's Kewpie Corporation introduced an AI-based ingredient inspection system that is used to inspect diced potatoes, chopped carrots, and other ingredients found in its baby food and potato salad products. This system replaces a manual inspection process to improve food quality and create a more employee-friendly environment (The Kewpie Group 2020).

Despite the strong interest of firms in adopting 4IR technologies, the adoption of most technologies will remain on a limited scale between 2020 and 2025. This is likely due to the limited understanding of firms, and the high capital investment costs associated with adopting such technologies. It is therefore critical for policy makers to build awareness of the long-term cost–benefits that 4IR technologies can bring to the agro-processing industry.

A global survey of executives (McKinsey & Company 2020) showed that the COVID-19 pandemic has led firms to accelerate the digitization of their customer and supply-chain interactions and their internal operations by 3–4 years. Firms that expect the adoption of 4IR technologies to be accelerated by COVID-19 believe that it is management's decision toward greater digitization that spurs this adoption. However, agro-processing firms in Azerbaijan assess that COVID-19 will have limited impact on the adoption of 4IR technologies in their industry (Figure 5). Two reasons could explain firms' relatively muted expectations on the impact of COVID-19 on 4IR

> **Figure 5: Perceptions on Impact of the COVID-19 Pandemic on Adoption of Industry 4.0 Technologies in the Agro-Processing Industry in Azerbaijan (%)**
>
> **Only 26% of employers believe that the COVID-19 pandemic has accelerated or will accelerate the use of 4IR technologies**
>
> Percent of surveyed firms
>
>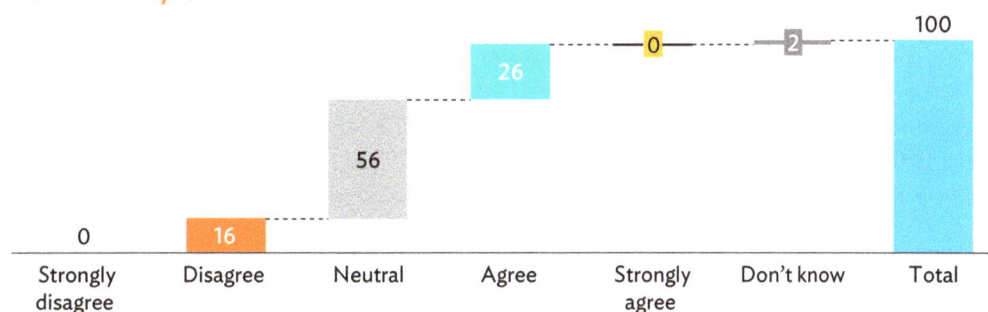
>
> **Common reasons for accelerated adoption:**
>
> Company management push for greater digitization in strategic shift
>
> Lack of labor due to movement restrictions necessitates more automation and shifting of activities to digital means
>
> Note: Based on survey of employers in the agro-processing industry between June and September 2021 (n=50).
> Source: Asian Development Bank (Sustainable Development and Climate Change Department).

technology adoption. First, firms have a limited understanding of 4IR technologies and their applications. Second, the oil-dependent Azerbaijani economy was severely impacted by the fall in global oil prices due to COVID-19, and tourism revenues were similarly affected. Unemployment rates increased sharply between 2019 and 2020.[5] This could create overall uncertainty among firms on their future economic prospects and therefore lead to unwillingness to invest in 4IR technologies.

Skills Demand Analysis

Job Implications

To determine the impact that the adoption of 4IR technologies will have on employment in Azerbaijan's agro-processing industry by 2025, the displacement and productivity effects of adopting 4IR technologies were estimated to determine the net change in job numbers (Figure 6).

(i) **Displacement effect.** This refers to the number of jobs that could potentially be lost due to automation using 4IR technologies. Around 30,000 jobs or 42% of the current workforce size in Azerbaijan's agro-processing industry could potentially be displaced due to the adoption of 4IR technologies in the next 5 years.

5 World Bank. Unemployment Data. https://data.worldbank.org/indicator/SL.UEM.TOTL.ZS?locations=AZ (accessed 15 September 2021).

(ii) **Productivity effect.** This refers to the job gains due to improved productivity from technology adoption, which increases the potential total output of the industry and the demand for labor. If policies that encourage full 4IR adoption are implemented, up to 45,000 jobs could be created in Azerbaijan's agro-processing industry by 2025 due to the productivity effect based on our modeling.

This report estimates that the jobs created by the productivity effect exceed those displaced to create close to 15,000 new jobs or the equivalent of 20% of the 2020 agro-processing workforce in net job gains by 2025. This is over and beyond the jobs created by BAU growth (Box 1).

In interpreting the gains from 4IR adoption, it is important to understand that the 15,000 net jobs estimated will be created over and beyond new jobs due to BAU growth. In other words, the agro-processing workforce grew at approximately 4.8% per annum[6] from 2015 and 2020, and if this growth were extrapolated up to 2025, an additional 19,000 jobs would have been created even without adopting 4IR technologies. However, the full adoption of 4IR technologies could see 34,000 more jobs in 2025 than in 2020.

Box 1: Estimating Employment Changes Due to Adoption of Industry 4.0 Technologies

To determine the impact that the adoption of Fourth Industrial Revolution (4IR or Industry 4.0) technologies will have on employment in Azerbaijan's agro-processing industry from 2020 to 2025, data on the business-as-usual (BAU) growth of the industry and responses from the June–September 2021 employers' survey were used. The displacement and productivity effects of adopting 4IR technologies were estimated to determine the net change in jobs.

Displacement effect. This refers to the number of jobs potentially lost due to automation using 4IR technologies. It assumes that as output per worker increases due to the adoption of 4IR technologies, fewer workers would be needed to produce the same amount of output under a BAU scenario in 2025. The BAU amount of output in 2025 was calculated based on growth rates before the coronavirus disease (COVID-19) pandemic struck, i.e., from 2014 to 2019, while industry labor productivity growth rates from 2014 to 2019 (before the COVID-19 pandemic) were used to calculate the BAU labor productivity (i.e., output per worker) in 2025. The expected labor productivity increase from the adoption of 4IR technologies, on top of the expected BAU labor productivity in 2025, was obtained from the employer survey to calculate the displacement effect.

Productivity effect. This refers to the job gains due to improved productivity from technology adoption, which increases the potential total output of the industry. For instance, the improved ease in production or the manufacturing of higher-quality goods can both lead to higher total industry output and corresponding job creation. This assumes that (i) the market can completely absorb the higher output produced; and (ii) firms can produce at the original cost notwithstanding the higher productivity (e.g., pay wages based on the BAU productivity of workers prior to 4IR adoption). Using the labor productivity increase estimated from the survey, on top of the labor productivity increase that would have taken place at BAU between 2020 and 2025, the industry's potential total output with 4IR adoption by 2025 was calculated, and the number of additional workers that could be hired if labor productivity had remained at BAU levels in 2025 was determined.

Net gains. The net gains were determined by taking the net of the increase in jobs created by the productivity impact and decrease in jobs created by the displacement impact.

Source: Asian Development Bank (Sustainable Development and Climate Change Department).

---

[6]   Calculated using industry and workforce data from State Statistical Committee of Azerbaijan. The sectoral growth rate figure was based on the food, beverage, and tobacco categories of the manufacturing sector.

Figure 6: Estimated Impact of Industry 4.0 on Number of Jobs by 2025 in the Agro-Processing Industry in Azerbaijan (%)

The adoption of 4IR technologies will lead to 20% more jobs in 5 years' time in the agro-processing sector

Percent of jobs impacted due to displacement and productivity effects of 4IR in 5 years' time

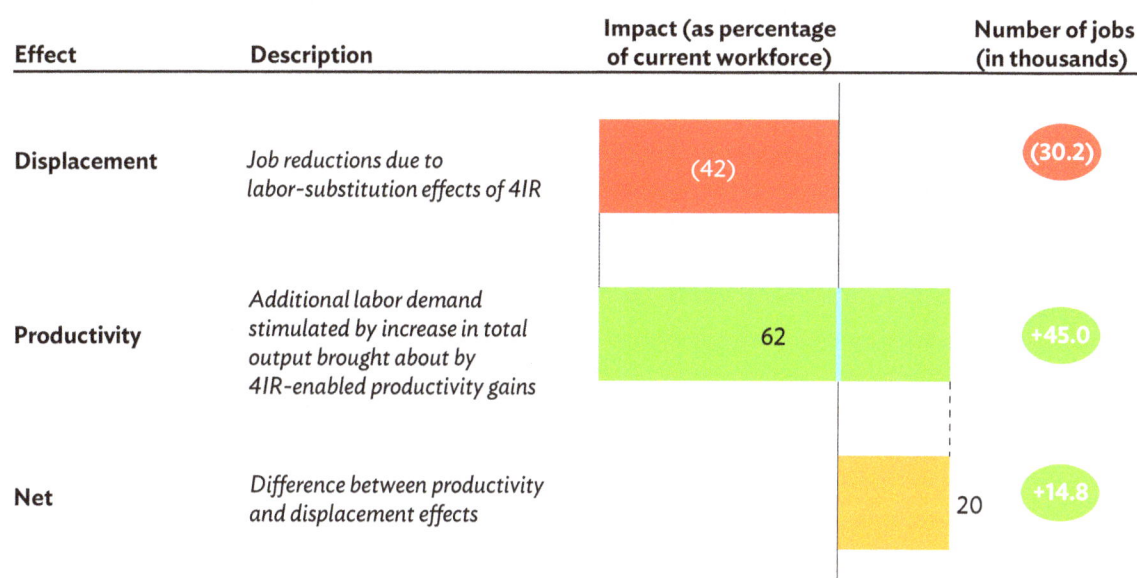

Effect	Description	Impact (as percentage of current workforce)	Number of jobs (in thousands)
Displacement	Job reductions due to labor-substitution effects of 4IR	(42)	(30.2)
Productivity	Additional labor demand stimulated by increase in total output brought about by 4IR-enabled productivity gains	62	+45.0
Net	Difference between productivity and displacement effects	20	+14.8

( ) = negative.
Note: Based on survey of employers in the agro-processing industry between June and September 2021 (n=50).
Sources: Industry employment data from State Statistical Committee of Azerbaijan. Labour Market. https://www.stat.gov.az/source/labour/?lang=en; industry output data from State Statistical Committee of Azerbaijan. Industry of Azerbaijan. https://www.stat.gov.az/source/industry/?lang=en.

There are however two caveats to realizing these job gains. First, these gains assume that firms in Azerbaijan fully adopt 4IR technologies. However, only 14% of firms in the agro-processing industry have a strong understanding of such technologies and their applications as of 2020. Second, new jobs created could be different from the jobs displaced, and based on the surveys conducted, would require different skill sets. As such, those who risk losing their jobs may not seamlessly move into new jobs being created without extensive reskilling.

To better understand the differences between the jobs displaced and the jobs created, it is important to recognize that each industry is characterized by various occupational roles, and the impact of 4IR technologies on these roles is uneven. For example, manual roles such as that of a basic food packing machine operator or cleaner are at a higher risk of automation, while the number of technical roles is more likely to increase as more workers are needed to operate and repair machines or use digital platforms. In this report, jobs in the agro-processing industry are split into five occupational groups (Table 2).

Table 2: Occupational Groups in the Agro-Processing Industry

	Occupational Group	Possible Job Titles
1	Technical	• Engineering Technician • Factory Process Control Technician
2	Managerial	• Chief Executive Officer • Factory Manager
3	Customer-facing	• Telephone Operator • Sales Assistant
4	Administrative	• Secretary • Finance Executive
5	Elementary and/or manual jobs	• Basic Food Packing Machine Operator • Cleaners

Sources: Asian Development Bank (Sustainable Development and Climate Change Department) and AlphaBeta.

Across all occupational groups, the employer surveys show that employers expect 4IR to decrease the number of workers as their jobs are replaced by automation (Figure 7). The decrease is particularly significant for manual workers and administrative workers. For manual workers, automated robots on the production floor could reduce the number of workers needed to cut raw materials or pack finished products. In the case of administrative workers, Big Data analytics could be used to calculate and predict sales and carry out demand forecasting, reducing the need for finance executives. However, 32% of employers also expect the number of technical

Figure 7: Employers' Expectations on Impact of Industry 4.0 on the Number of Jobs by 2025 in the Agro-Processing Industry in Azerbaijan (%)

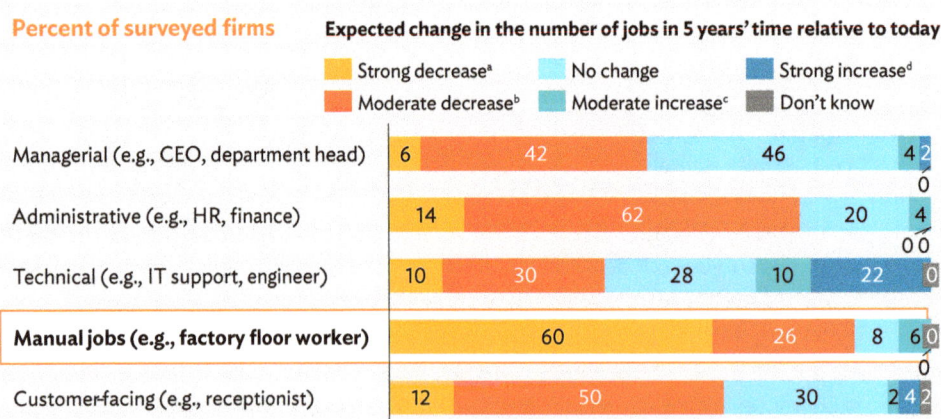

In the agro-processing industry, manual jobs are expected to decrease most significantly due 4IR technologies adoption

CEO = chief executive officer, HR = human resources, IT = information technology.
a Greater than or equal to 50% decrease in number of jobs.
b Less than 50% decrease in number of jobs.
c Less than 50% increase in number of jobs.
d Greater than or equal to 50% increase in number of jobs.
Note: Based on survey of employers in the agro-processing industry between June and September 2021 (n=50).
Source: Asian Development Bank (Sustainable Development and Climate Change Department).

workers to increase as more technically trained workers are needed to operate machines and data monitoring and collection systems.

These shifts mean that the workforce of the agro-processing industry will look very different by 2025 (Figure 8). The proportion of manual jobs will decrease by over 7 percentage points, from 33.9% to 26.8% while the proportion of technical jobs will correspondingly increase. This implies that workers displaced by 4IR technologies are likely to be from manual roles that often require lower levels of education and skills and pay less, while jobs created are likely to be in better-paid technical roles that require higher levels of education. On one hand, it means that better jobs will be available to Azerbaijani workers across the agro-processing industry. Jobs created could also be safer and more satisfying for workers. For instance, apple sorting and packing could lead to hand and wrist injuries for workers as well as neck and back injuries from staying in the same position for prolonged periods (Department of Environmental Health 2001). The robotic crate packing solution used by the UK fruit packer Adrian Scripps to pack apples could potentially alleviate these issues and improve workplace safety for workers (South East Farmer 2021). On the other hand, it means that (i) strong social protection mechanisms and reskilling policies are needed to ensure that manual workers are not adversely impacted by the adoption of 4IR technologies; and that (ii) without a skilled workforce to support the adoption of 4IR technologies, firms cannot complete the transition toward 4IR.

Apart from ensuring that less educated manual workers can benefit from the shift toward 4IR, it is equally important that women are not left behind in these gains. Based on the employer surveys conducted, manual and customer-facing roles in the agro-processing industry have a higher proportion of women compared to other job roles. As such, with most of the job gains expected in technical roles, it is unsurprising that men will receive 1.6 times more of the net job gains from 4IR adoption compared to female workers (Figure 9). Reskilling policies to ensure that manual or administrative workers displaced by automation can transit into new, technical roles must place special focus on ensuring that female workers can also benefit from 4IR. International agencies

Figure 8: Composition of Jobs in 2020 and by 2025 in the Agro-Processing Industry in Azerbaijan (%)

The distribution of jobs will change—manual jobs will record the largest fall while technical jobs will see the largest increase in 5 years' time

Weighted average percentage share of employees by occupational group in surveyed firms[a]

● Negative shift
● Positive shift

Occupational group	Share today	Share in 5 years' time[b]	Percentage shift
Manual jobs	33.9	26.8	(7.1%)
Administrative	11.9	11.6	(0.3%)
Customer-facing	14.1	14.6	+0.5%
Managerial	12.5	13.5	+1.0%
Technical	27.5	33.7	+6.2%

( ) = negative.

Note: Based on survey of employers in the agro-processing industry between June and September 2021 (n=50).

[a] Average share of employees in surveyed firms is weighted by the number of employees in each firm, as indicated by respondents; percentages might not add up to 100% due to rounding.

[b] The change in the number of workers in each job type is based on the number of firms indicating different levels of changes in number of jobs, i.e., "strong increase," "moderate increase," "no change," "moderate decrease," and "strong decrease." The midpoint of the range for each option for expected change is used; for expected increase or decrease of over 50%, we used the low-bound of 50% was used.

Source: Asian Development Bank (Sustainable Development and Climate Change Department).

Figure 9: Estimated Net Job Gains by Gender from Industry 4.0 Adoption by 2025 in the Agro-Processing Industry in Azerbaijan ('000s)

The net gains in jobs will benefit male workers more than female workers, implying that policies are needed to ensure that adoption benefits are equitable

Estimated number of net jobs created by gender (in thousands)

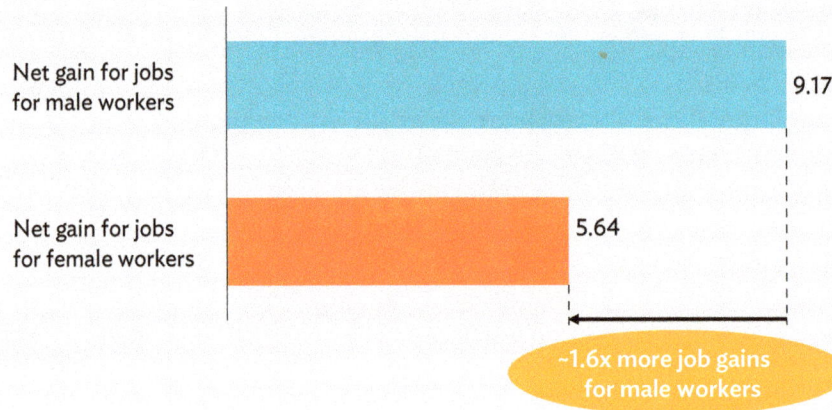

Net gain for jobs for male workers — 9.17

Net gain for jobs for female workers — 5.64

~1.6x more job gains for male workers

Note: Based on survey of employers in the agro-processing industry between June and September 2021 (n=50).
Source: Industry employment data from State Statistical Committee of Azerbaijan. Labour Market. https://www.stat.gov.az/source/labour/?lang=en.

such as the Asian Development Bank (ADB) and International Labour Organization (ILO) could lend their financial and technical support in this area. In Indonesia, the Philippines, and Thailand, ILO aims to enhance the employability of women for jobs in science, technology, engineering, and mathematics (STEM) through its Women in STEM Workforce Readiness and Development Program. The program provides training in critical soft and technical STEM-related skills, as well as targeted mentorship opportunities. Participants benefit from tangible employment opportunities upon the completion of the program, which includes activities to identify skills gaps, skill training programs, job placement schemes, and in-company developing and mentoring (ILO 2020a).

Task Implications

To understand the impact of the adoption of 4IR technologies on jobs and skills demand, it is important to understand that technology does not automate jobs, but rather individual tasks or combinations of tasks. This report examines five types of tasks linked to jobs in the agro-processing industry and how they could be impacted by 4IR:

(i) **Routine physical.** These tasks involve repetitive and predictable physical work, for example, a factory worker packing food products on a manufacturing line.

(ii) **Routine interpersonal.** These tasks involve predictable interactions with other people, for example, a customer sales agent reading a sales script.

(iii) **Nonroutine physical.** These tasks involve physical work that is not repetitive or predictable, for example, a mechanic diagnosing and repairing factory equipment.

(iv)   **Nonroutine interpersonal.** These tasks involve complex or creative interactions with other people, for example, supervising others or making speeches or presentations.

(v)   **Analytical.** These are tasks that vary significantly and involve a strong thinking and analytical component. They predominantly involve computers or other technological equipment.

Employers were asked to estimate how much time an average employee spent on each task in an average work week as of 2020 (based on 2021 survey data), and to predict how that would change by 2025 with the adoption of 4IR technologies. The analysis reveals that employers expect to see a sharp drop in the amount of time workers spend on routine, physical tasks, and an increase in the amount of time spent on analytical tasks (Figure 10). While 46% of weekly working hours is spent on routine tasks in 2020, only 39.5% of weekly working hours is expected to be spent on routine tasks by 2025. This could lead to increased job satisfaction among workers as machines take over a greater share of full routine tasks, with past research showing that monotonous, automatable tasks are the least satisfying tasks to perform (AlphaBeta 2017). These findings are consistent with the expected shifts in the distribution of jobs across different occupational groups. The number of manual workers (i.e., workers most likely to be engaged in routine, physical tasks) will decrease while the number of technical workers (i.e., workers most likely to be engaged in analytical tasks) will increase.

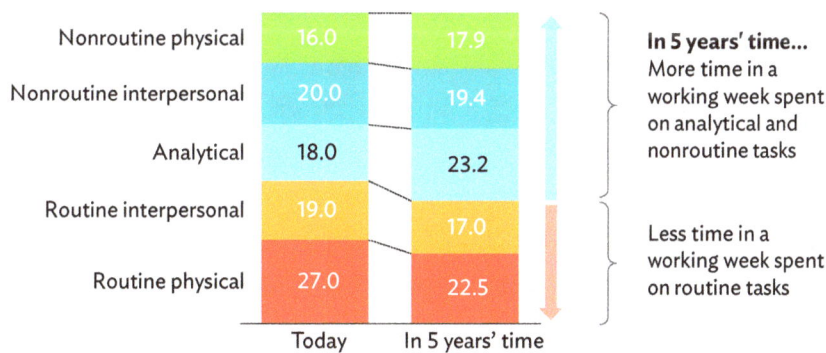

Figure 10: Time Spent by Employees on Tasks at Work in 2020 and by 2025 in the Agro-Processing Industry in Azerbaijan (%)

Adoption of 4IR technologies may shift the overall distribution of weekly working hours from routine to analytical and nonroutine tasks

Average percentage share of weekly working hours spent by task in surveyed firms

Task	Today	In 5 years' time
Nonroutine physical	16.0	17.9
Nonroutine interpersonal	20.0	19.4
Analytical	18.0	23.2
Routine interpersonal	19.0	17.0
Routine physical	27.0	22.5

In 5 years' time...
More time in a working week spent on analytical and nonroutine tasks

Less time in a working week spent on routine tasks

Note: Based on survey of employers in the agro-processing industry between June and September 2021 (n=50).
Figures include rounding adjustments.
Source: Asian Development Bank (Sustainable Development and Climate Change Department).

Skills Implications

These task shifts will impact the skills required in the industry. While employers indicated only slight shifts in the proportion of time allotted to each type of task, the skills that are valued by employers in the agro-processing industry will see a change in relative importance due to these shifts. This analysis considers 10 categories of skills as set out in Table 3.

Table 3: Categories of Skills Considered in the Analysis

No.	Skill	Definition
1	Creative thinking and/or design	Ability to develop, design, or creating new applications, ideas, relationships, systems, or products
2	Critical thinking	Ability to use logic and reasoning to identify the strengths and weaknesses of alternative solutions, conclusions or approaches to problems
3	Adaptive learning	Ability to pick up new skills as demanded by the job
4	Complex problem solving	Ability to identify complex problems and review related information to develop and evaluate options and implement solutions
5	Digital and/or information and communication technology skills	Ability to design, set up, operate, and correct malfunctions involving application of machines or technological systems
6	Numeracy	Ability to add, subtract, multiply, or divide quickly and correctly and use mathematics to solve problems
7	Written Communication	Ability to read and understand information and ideas presented in writing, and to communicate information and ideas in writing
8	Verbal Communication	Ability to communicate information and ideas clearly by talking to others
9	Management	Ability to motivate, develop, and direct people as they work, and to identify the best people for the job
10	Social and interpersonal	Ability to work with people to achieve goals

Source: Asian Development Bank (Sustainable Development and Climate Change Department).

This analysis reveals that alongside the change in tasks, the adoption of 4IR technologies will change the types of skills valued by employers. In 2020, employers rank critical thinking as the most important skill followed by written communication and digital or ICT skills. As the ranking was derived from an analysis of online job listings, it is likely that these are skewed toward employers in higher-level roles and not indicative of the skills required of a manual worker at present. Nevertheless, the employer survey reveals that digital and/or ICT skills and creative thinking and/or design skills will become more important to employers for all categories of workers by 2025 (Figure 11). In the agro-processing context, an increased demand for creative thinking and/or design skills could imply a greater demand for workers able to carry out digital marketing campaigns or find creative applications for additive manufacturing. For instance, researchers at Redefine Meat are applying 3D printing technology, meat digital modeling, and advanced food formulations to produce animal-free meat with the appearance, texture, and flavor of whole muscle meat (Channel News Asia 2021). Among the skills prioritized by employers by 2025, critical thinking and digital and/or ICT skills are areas in which employers also see significant proficiency gaps (Figure 12). Of the employers who feel that a step-up from basic proficiency in critical thinking skills is needed, 37% feel that a step-up to the intermediate level of proficiency is needed while 63% would like to see a step-up to advanced proficiency.

Figure 11: Importance of Skills in 2020 and for Industry 4.0 Adoption by 2025 in the Agro-Processing Industry in Azerbaijan

There is a significant change in skills perceived as important by employers over the next 5 years

Skills of increasing importance in 5 years' time
Skills of decreasing importance in 5 years' time
Skills with no change in importance in 5 years' time

Importance ranking	Today[a]	In 5 years' time[b]	Change in ranking
1	Critical thinking	Digital and/or ICT skills	+2
2	Written communication	Creative thinking or design	+6
3	Digital and/or ICT skills	Critical thinking	-2
4	Complex problem solving	Social and interpersonal	+3
5	Verbal communication	Numeracy	+4
6	Management	Management	-
7	Social and interpersonal	Adaptive learning	+3
8	Creative thinking or design	Written communication	-6
9	Numeracy	Complex problem solving	-5
10	Adaptive learning	Verbal communication	-5

— = not applicable, ICT = information and communication technology.
[a] Evaluated using the employer survey and supported by job portal data.
[b] Evaluated using the employer survey.
Note: Based on survey of employers in the agro-processing industry between June and September 2021 (n=50); job portal data on the agro-processing industry from the job portal Offer.AZ (accessed June 2021).
Source: Asian Development Bank (Sustainable Development and Climate Change Department).

Figure 12: Required Step-Up in Employees' Level of Skill Proficiency from Today for Industry 4.0 Adoption by 2025 in the Agro-Processing Industry in Azerbaijan (%)

To be 4IR-ready, workers would require proficiency leaps in creative thinking as well as written and verbal communication skills

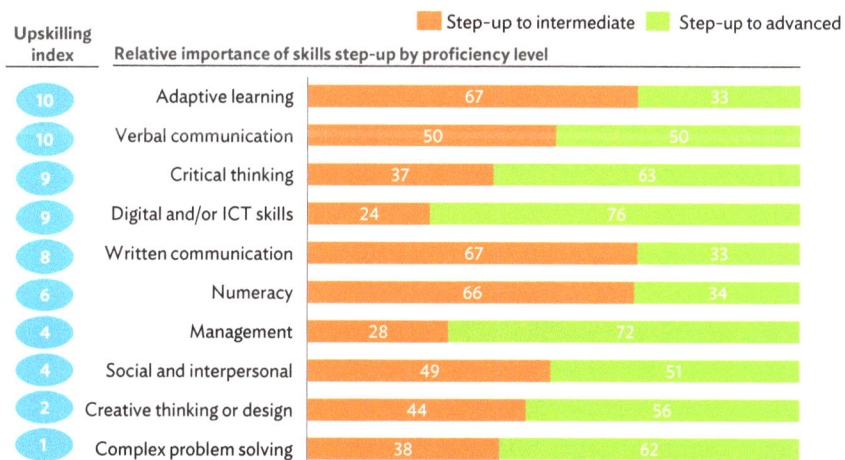

Step-up to intermediate   Step-up to advanced

Upskilling index	Relative importance of skills step-up by proficiency level		
10	Adaptive learning	67	33
10	Verbal communication	50	50
9	Critical thinking	37	63
9	Digital and/or ICT skills	24	76
8	Written communication	67	33
6	Numeracy	66	34
4	Management	28	72
4	Social and interpersonal	49	51
2	Creative thinking or design	44	56
1	Complex problem solving	38	62

4IR = Fourth Industrial Revolution, ICT = information and communication technology.
Note: Based on survey of employers in the agro-processing industry between June and September 2021 (n=50).
Source: Asian Development Bank (Sustainable Development and Climate Change Department).

(i)   Skills Supply Trends

Overall, agro-processing businesses in Azerbaijan appear to face challenges in recruiting workers with the requisite skill sets. Close to half of firms surveyed disagree that it is easy to identify and recruit high-quality graduates for entry-level positions (Figure 13). Only 6% of firms strongly agree and 28% of firms agree that graduates hired over the previous year have the appropriate general skills to be effective in entry-level positions. Past surveys and consultations with industry stakeholders similarly point to employers in Azerbaijan having difficulties in recruiting candidates with the appropriate "soft skills" such as social and interpersonal skills. A Business Climate Survey released by the German-Azerbaijani Chamber of Commerce (AHK Azerbaijan) and KPMG Azerbaijan in 2020 found the lack of a highly skilled local workforce is one of the top five operational and economic challenges faced by foreign businesses in Azerbaijan (German-Azerbaijani Chamber of Commerce (AHK Azerbaijan) and KPMG Azerbaijan 2020). In the same vein, firms also appear to face challenges in providing hired workers with the requisite training. Only 32% of firms agree that they current invest sufficiently in training employees (Figure 14). The lack of sufficient investment in training could be in part due to the lack of good trainers available. Over 60% of firms disagree that it is easy to find good quality trainers. Policies or programs to enable industry professionals to learn pedagogical approaches and transit into trainer provider roles could help to address this shortage of good trainers.

To understand the types of training channels that employers relied on to address the skills shortage, four types of training channels were examined in relation to how they would be tapped to provide skills training for employees in 2020 and going forward (Table 4).

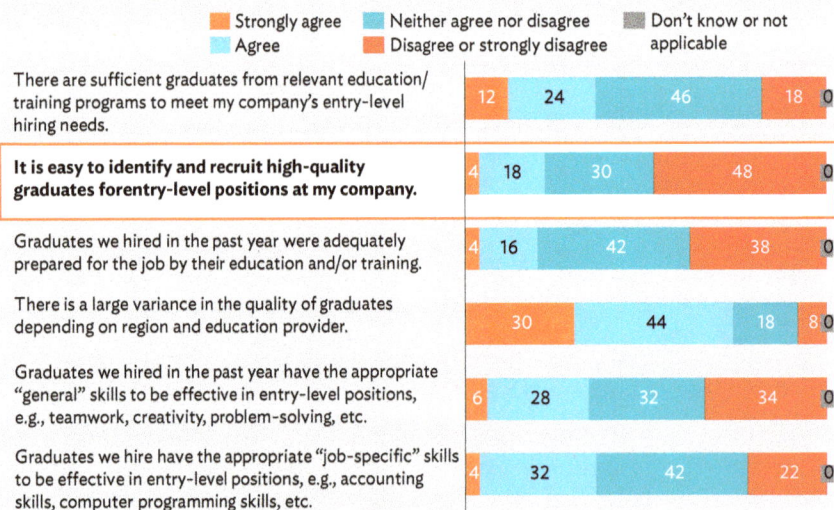

Figure 13: Employer Sentiment Toward Graduates Hired in the Agro-Processing Industry in Azerbaijan (%)

Close to half of employers disagree that it is easy to identify and recruit high-quality graduates for entry-level positions

Percent of surveyed firms

Legend: Strongly agree | Neither agree nor disagree | Don't know or not applicable | Agree | Disagree or strongly disagree

Statement	Strongly agree	Agree	Neither agree nor disagree	Disagree or strongly disagree	Don't know or not applicable
There are sufficient graduates from relevant education/training programs to meet my company's entry-level hiring needs.		12	24	46	18	0
It is easy to identify and recruit high-quality graduates forentry-level positions at my company.	4	18	30	48	0
Graduates we hired in the past year were adequately prepared for the job by their education and/or training.	4	16	42	38	0
There is a large variance in the quality of graduates depending on region and education provider.	30	44	18	8	0
Graduates we hired in the past year have the appropriate "general" skills to be effective in entry-level positions, e.g., teamwork, creativity, problem-solving, etc.	6	28	32	34	0
Graduates we hire have the appropriate "job-specific" skills to be effective in entry-level positions, e.g., accounting skills, computer programming skills, etc.	4	32	42	22	0

Note: Based on survey of employers in the agro-processing industry between June and September 2021 (n=50).
Source: Asian Development Bank (Sustainable Development and Climate Change Department).

Figure 14: Employers' Perception on Training for Employees in the Agro-Processing Industry in Azerbaijan (%)

A total of 64% of employers find it difficult to find good quality training providers to train their employees

Percent of surveyed firms

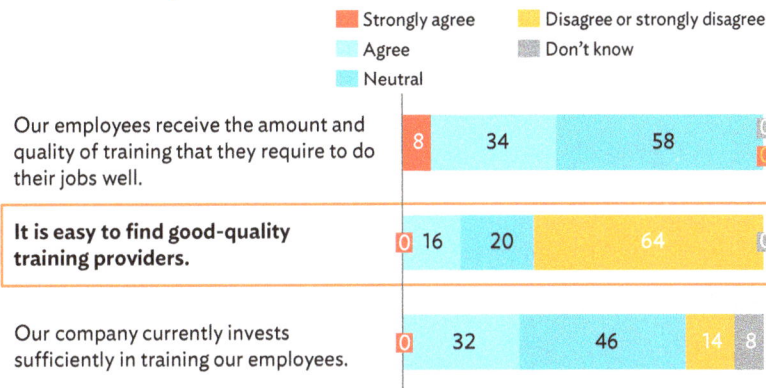

Legend: Strongly agree | Disagree or strongly disagree | Agree | Don't know | Neutral

Statement	Strongly agree	Agree	Neutral	Disagree or strongly disagree	Don't know
Our employees receive the amount and quality of training that they require to do their jobs well.	8	34	58	0	0
It is easy to find good-quality training providers.	0	16	20	64	0
Our company currently invests sufficiently in training our employees.	0	32	46	14	8

Note: Based on survey of employers in the agro-processing industry between June and September 2021 (n=50).
Source: Asian Development Bank (Sustainable Development and Climate Change Department).

Table 4: Four Types of Training Channels

Training Channel	Description
1 **On-the-job training**	Training that takes place within the company as the employee performs the actual work. These are typically provided by a more senior or experienced coworker and can also be in the form of internally organized sessions conducted by coworkers.
2 **Flexible online training**	Online courses that are subsidized or sponsored by the company for their employees (e.g., courses from online training platforms like Udemy and Coursera). Such online training courses tend to be flexible in terms of when workers may access the content, and typically allow workers to gain industry-recognized micro-credentials or micro-degrees at the end of the course.
3 **Professional courses**	Short courses that are sponsored or organized by the company for their employees. These are conducted by professional instructors and are typically held within contained periods spanning at least 1 week and up to 6 months.
4 **Formal education courses**	Such courses are those taken at higher education or technical and vocational education and training institutions that are subsidized or sponsored by the company for their employees. Such courses tend to be specially designed for working professionals, such as part-time diplomas or master's degrees.

Source: Asian Development Bank (Sustainable Development and Climate Change Department).

Less than half of employers agree that their employees receive a sufficient amount and quality of training to do their jobs well in 2020 (Figure 14). However, employers expect the amount of training provided to employees to fall across all four training channels identified by 2025 (Figure 15). This could be due to the resource crunch and uncertainty on future economic conditions created by the COVID-19 pandemic, so that firms are unwilling to spend more resources to train their employees. This suggests that government support to provide training incentives or subsidies would be needed to help firms prepare their workforce for 4IR. During the country consultations, local experts and stakeholders highlighted that while companies currently provide workplace training or industry apprenticeships, the skills acquired by trainees are not always recognized. Efforts were ongoing to strengthen the structure of workplace-based training and improve the recognition of skills acquired through nonformal training. These would require strong collaboration between industry stakeholders and training providers. For instance, in the UK, apprenticeship standards are developed by an employer group (the Institute for Apprenticeships) under the sponsorship of the Department of Education. The apprenticeship standard sets out the skills, knowledge, and behaviors required of a qualified worker, and a separate document sets out how these are to be assessed at the end of the apprenticeship program (ILO 2020b).

Figure 15: Proportion of Employees Receiving Training in 2020 and Requiring Training by 2025 for Each Training Channel in the Agro-Processing Industry in Azerbaijan (%)

Employers in Azerbaijan expect the proportion of employees receiving training to fall across various channels in five years' time

Percentage share of employees by training channel

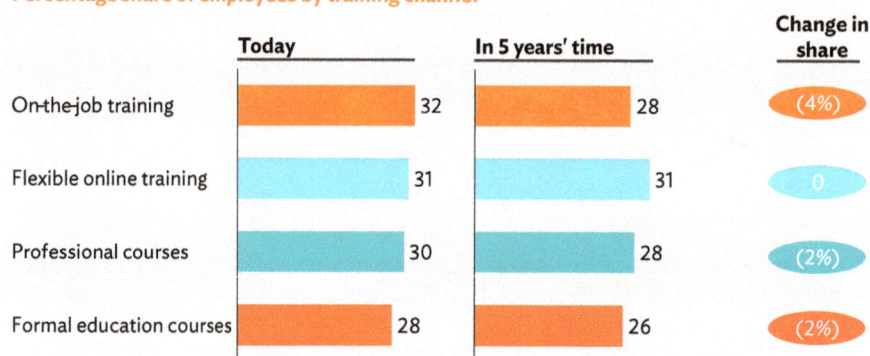

	Today	In 5 years' time	Change in share
On-the-job training	32	28	(4%)
Flexible online training	31	31	0
Professional courses	30	28	(2%)
Formal education courses	28	26	(2%)

( ) = negative.

Note: Based on survey of employers in the agro-processing industry between June and September 2021 (n=50). The sum of all shares for 2020 and by 2025 exceed 100%, as one employee can undergo training in more than one training channel.

Source: Asian Development Bank (Sustainable Development and Climate Change Department).

D.   Transportation and Storage Industry

Like in the agro-processing industry, most firms in the transportation and storage industry have a limited understanding of 4IR technologies. The digitization of the supply chain, or supply chain 4.0, has strong potential to improve the efficiency, accuracy, and speed of transportation and storage operations and it is critical to build awareness of these benefits among firms in Azerbaijan (McKinsey & Company 2016).

If awareness on 4IR technologies is higher among transportation and storage firms in Azerbaijan and full adoption is achieved in the industry, net job gains from 4IR can be expected. Around 27,000 new jobs or the equivalent of 13% of the 2020 transportation and storage workforce are expected to be created by 2025. This gain is over and beyond the BAU growth of the industry's labor force. The distribution of jobs will also shift, where the proportion of manual jobs will decrease while the share of technical jobs will increase. In tandem with this shift, creative thinking and/or design, as well as numeracy skills, will become more important to employers by 2025 as reported by the employer surveys. This significant shift in the type of roles and required skills implies there will be a need for a strong focus on worker training, although employers in the industry highlighted some worrying trends about the adequacy of existing training programs and gaps into their own training efforts.

Relevance to Industry 4.0

Previous research has identified a range of 4IR technologies critical to the transformation of the transportation and storage industry. This includes IOT technology to track products along the entire supply chain and integrated logistics systems that link various parts of the transportation process.

Some key technologies include the following:

(i) **Autonomous robots.** Autonomous robots are intelligent machines capable of performing tasks with a high degree of autonomy. Autonomous robots can be deployed to warehouses to move, sort, and find parcels as well as track inventory. Autonomous vehicles and drones can also be deployed in for autonomous driving and last-mile delivery of goods (Banker 2020). Daimler's Highway Pilot System increases road safety by relieving the driver during potentially dangerous sections of a route. Daimler's research found that assistance from onboard technology could reduce driver sleepiness by 25% compared with driving a conventional truck and reduce accidents caused by human error (AIG 2017).

(ii) **Artificial intelligence.** The AI technology gives machines the ability to learn and act intelligently and carry out a wide range of human-like processes. This means they can make decisions, carry out tasks, and even predict future outcomes based on what they learn. In the transportation and storage industry, AI technologies can be used to draw information from various variables, such as traffic patterns and weather patterns, to allow dynamic route optimization and predict delivery timings. [7] Artificial intelligence can also be used for intelligent optical character recognition programs that read both printed and handwritten text can also be used to streamline paperwork as well as in chatbots (footnote 9). Courier company UPS uses an online platform called Network Planning Tools, which allows its engineers to view activity at UPS facilities around the world and route shipments to the ones with the most capacity. The Network Planning Tools uses AI to create forecasts about package volume and weight based on analysis of historical data, and machine-learning algorithms to analyze decisions the company's engineers made and assess how they affected customer satisfaction and internal costs. Through identifying and eliminating bottlenecks, UPS estimates that the program can save the company from $100 million to $200 million a year (*MIT Technology Review* 2018). UPS also uses an AI-enabled chatbot, the UPS Bot, which mimics human conversation and can respond to customer queries, track packages, and give out shipping rates (Marr 2018).

(iii) **Internet of Things.** The IOT is a network of sensors and actuators embedded in machines and other physical objects that connect with one another and the internet. It has a wide range of applications, including data collection, monitoring, decision-making, and process optimization (McKinsey Global Institute 2014). IOT sensors can be used to gauge inventory and track products along the entire supply chain. They are used to keep facilities secure and monitor variables such as light and heat in warehouses as well as for fleet management or damage monitoring.[8] DHL Supply Chain partnered with clean-tech start-up BeeBryte to leverage its smart heating, ventilation, and air-conditioning unit for their

---

[7] DHL. Artificial Intelligence: AI. Today a Novelty, Tomorrow a Necessity. https://www.dhl.com/sg-en/home/insights-and-innovation/insights/artificial-intelligence.html.

[8] DHL. Internet of Things. https://www.dhl.com/global-en/home/insights-and-innovation/thought-leadership/trend-reports/internet-of-things-in-logistics.html.

Singapore facility. DHL benefited from 40% energy cost savings through real-time adjustments, based on anticipated weather conditions, building occupancy, and business activity, to maintain its facility temperature within a preferred operating range (footnote 10).

(iv) **Blockchain.** Blockchain is a shared, immutable ledger that facilitates the process of recording transactions and tracking assets in a business network.[9] Blockchain technology can be used to track events at all points of the supply chain to improve accountability and reduce fraud or error. Digital certificates and tamper-proof documents can be used to improve efficiency in global trade by reducing bureaucracy and paperwork (DHL 2018). For instance, to unlock efficiency in ocean freight, Maersk and IBM established a global blockchain-based system for digitizing trade workflows and end-to-end shipment tracking (Groenfeldt 2017). The system allows each stakeholder in the supply chain to view the progress of goods and see the status of customs documents and other data. Blockchain technology ensures secure data exchange (IBM 2018). The digitization of the workflow and reduced paperwork is estimated to save around $300 per container in shipping costs or over $5 million for a large container ship (*Ship Technology* 2017).

(v) **Systems integration.** Systems integration technologies link different computing systems and software applications to act as a coordinated whole. An integrated logistics systems that links various systems including sales, warehousing, freight, as well last-mile fulfillment can contribute to overall efficiency and higher customer satisfaction (Singapore Institute of Purchasing and Materials Management 2019). For example, digital twin systems are used in warehouses and port facilities to consolidate comprehensive data on the movement of inventory, equipment, and personnel that can aid the efficiency of operations.[10] Communications company Ericsson's Maritime ICT Cloud platform connects vessels at sea to shore-based operations and harmonizes the flow of information between maintenance service providers, customer support centers, fleet and transport partners, port operations, and authorities. Sensors monitor everything from vessel location and speed to the status and temperature of refrigerated cargo containers, giving shipping firms and producers real-time information on their goods. The platform allows shipping firms to make their ships safer and less expensive through real-time data analysis of potential dangers and inefficiencies (Telecoms 2015).

About 50 transportation and storage firms were surveyed for the Azerbaijan employer surveys. Like the agro-processing industry, firms in the transportation and storage industry demonstrated a limited understanding of 4IR technologies. Only 20% of firms have a good understanding of 4IR technologies and their applications and 44% had not heard of 4IR before (Figure 16). Similarly, only 12% of transportation and storage firms believe that firms in their supply chain understand 4IR technologies well while 40% believe that firms in the supply chain had not heard of 4IR (Figure 17). This suggests that the concept of 4IR is relatively new to firms in Azerbaijan and policy makers would need to build awareness of the tools available and their benefits across various industries.

As with firms in the agro-processing industry, transportation and storage firms have strong expectations of the potential labor productivity gains from adopting 4IR technologies despite their limited understanding. On average, firms expect output per worker to increase by 41% by 2025 with the adoption of 4IR technologies (Figure 18).

In 2020, the adoption of 4IR technologies is limited among transportation and storage firms in Azerbaijan. However, adoption rates are likely to increase significantly across various 4IR technologies identified to be crucial to the industry with IOT, blockchain, and systems integration technologies seeing the largest increase in adoption levels (Figure 19).

---

[9]   IBM. What is Blockchain Technology? https://www.ibm.com/sg-en/topics/what-is-blockchain.

[10]   DHL. Digital Twins. https://www.dhl.com/global-en/home/insights-and-innovation/thought-leadership/trend-reports/virtual-reality-digital-twins.html.

Figure 16: Understanding of Industry 4.0 Technologies in the Transportation and Storage Industry in Azerbaijan (%)

Only 20% of employers in the transportation and storage industry have a good understanding of 4IR technologies and their applications

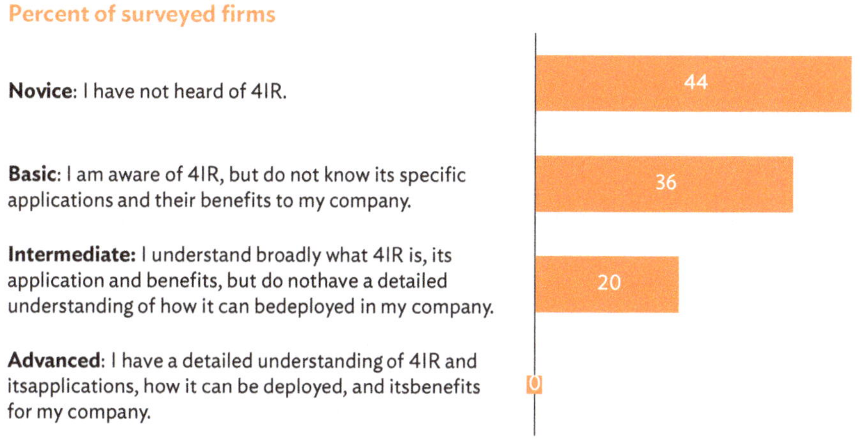

Percent of surveyed firms

Novice: I have not heard of 4IR. — 44

Basic: I am aware of 4IR, but do not know its specific applications and their benefits to my company. — 36

Intermediate: I understand broadly what 4IR is, its application and benefits, but do nothave a detailed understanding of how it can bedeployed in my company. — 20

Advanced: I have a detailed understanding of 4IR and itsapplications, how it can be deployed, and itsbenefits for my company. — 0

4IR = Fourth Industrial Revolution.
Note: Based on survey of employers in the transportation and storage industry between June and September 2021 (n=50).
Source: Asian Development Bank (Sustainable Development and Climate Change Department).

Figure 17: Understanding of Industry 4.0 Technologies Among Firms in the Transportation and Storage Supply Chain in Azerbaijan (%)

A total of 56% of transportation and storage firms believe that companies in their supply chain have a limited understanding of 4IR technologies

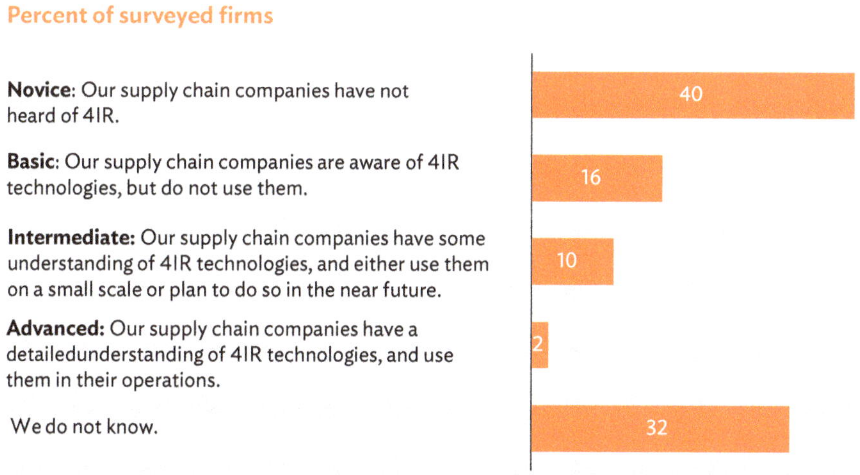

Percent of surveyed firms

Novice: Our supply chain companies have not heard of 4IR. — 40

Basic: Our supply chain companies are aware of 4IR technologies, but do not use them. — 16

Intermediate: Our supply chain companies have some understanding of 4IR technologies, and either use them on a small scale or plan to do so in the near future. — 10

Advanced: Our supply chain companies have a detailedunderstanding of 4IR technologies, and use them in their operations. — 2

We do not know. — 32

4IR = Fourth Industrial Revolution.
Note: Based on survey of employers in the transportation and storage industry between June and September 2021 (n=50).
Source: Asian Development Bank (Sustainable Development and Climate Change Department).

Figure 18: Expected Increase in Output per Worker Due to Industry 4.0 Technologies by 2025 in the Transportation and Storage Industry in Azerbaijan (%)

The majority of firms expect labor productivity to increase by at least 10% in 5 years' time with the adoption of 4IR technologies

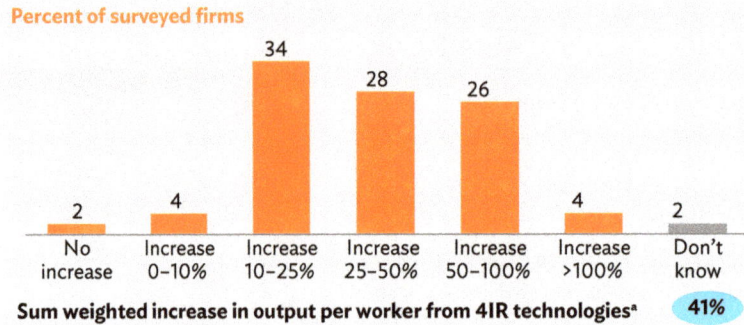

Percent of surveyed firms

No increase	Increase 0–10%	Increase 10–25%	Increase 25–50%	Increase 50–100%	Increase >100%	Don't know
2	4	34	28	26	4	2

Sum weighted increase in output per worker from 4IR technologies[a]   41%

4IR = Fourth Industrial Revolution.

Notes: Based on survey of employers in the transportation and storage industry between June and September 2021 (n=50). Calculated using sum-weighted average of output increase by the number of firms indicating different levels of expected increase in output, i.e., 0%, 0%–10%, 10%–25%, 25%–50%, 50%–100%, and over 100%. The midpoint of the range for each option for expected increase in output is used; for expected output increase of over 100%, the lower bound of 100% is used.

Source: Asian Development Bank (Sustainable Development and Climate Change Department).

Figure 19: Adoption of Relevant Industry 4.0 Technologies in the Transportation and Storage Industry in Azerbaijan (%)

The use of 4IR technologies is expected to increase in the transportation and storage industry over the next 5 years

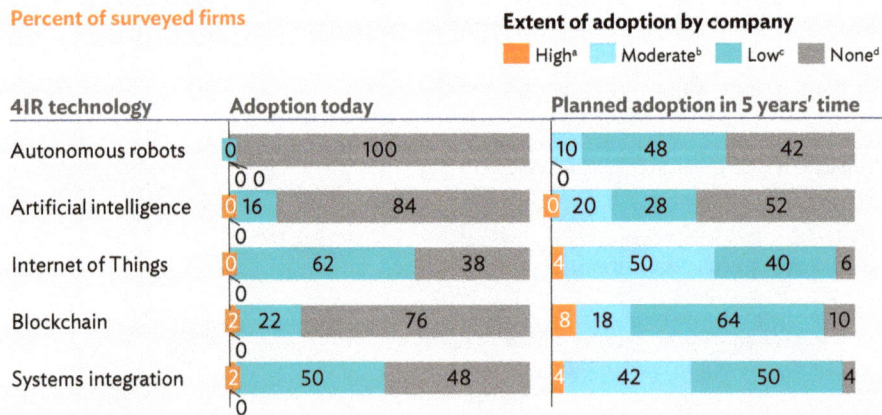

Percent of surveyed firms

Extent of adoption by company
High[a]   Moderate[b]   Low[c]   None[d]

4IR technology	Adoption today				Planned adoption in 5 years' time			
Autonomous robots	0		100		10	48		42
Artificial intelligence	0	16	84		0	20	28	52
Internet of Things	0	62		38	4	50		40	6
Blockchain	2	22	76		8	18	64	10
Systems integration	2	50	48		4	42	50	4

4IR = Fourth Industrial Revolution.

Notes: Based on survey of employers in the transportation and storage industry between June and September 2021 (n=50).

[a] "High": Firm has fully deployed the technology across all possible functions in the enterprise and/or has plans to fully deploy the technology across all possible functions in the future.

[b] "Moderate": Firm has implemented the technology, but not fully deployed across all possible functions in the enterprise and/or plans to implement the technology across a few functions in the future.

[c] "Low": Firm is experimenting with the technology at a very limited scale within the enterprise and/or plans to experiment with the technology in the future.

[d] "None": Firm has not used technology at all within the enterprise and/or has no plans to use the technology in the future.

Source: Asian Development Bank (Sustainable Development and Climate Change Department).

Currently, most firms either do not adopt the identified 4IR technologies or adopt them only to a limited extent. All firms surveyed do not implement autonomous robot technologies currently. This could be due to the relatively high costs of adopting such technologies and the challenges of incorporating advanced robotics technologies into existing work environments. For instance, retail chain Walmart uses autonomous robots to fetch items such as boxed and frozen food for online orders (Versai 2021). However, firms would not be able to do this without first ensuring that warehouses are organized in a way that is conducive and safe for such robots to operate.

Instead, firms are most keen on the adoption of IOT, blockchain, and systems integration technologies. IOT technologies are adopted by 62% of firms currently and more than 90% of firms expect to adopt them by 2025. In the transportation and storage industry, IOT-enabled devices can track products along the entire supply chain as well as keep facilities secure and regulate variables such as temperature, heat, and humidity in warehouses so that products remain unspoiled. The use of IOT technologies not only can increase labor productivity gains in the industry, but also help the industry move toward transporting higher value-add products such a medical supplies or vaccines that would need to be moved at specific temperatures (IEEE 2021).

Only 24% of transportation and storage firms believed that the COVID-19 pandemic will accelerate the use of 4IR technologies (Figure 20). Like agro-processing firms, this could be due to the resource crunch and general economic uncertainty created by the COVID-19 pandemic, so that firms are unwilling to invest significant resources in new technologies.

Figure 20: Perception on the Impact of the COVID-19 Pandemic on the Adoption of Industry 4.0 Technologies in the Transportation and Storage Industry in Azerbaijan (%)

Only 24% of employers believe that the COVID-19 pandemic has accelerated or will accelerate the use of 4IR technologies

Percent of surveyed firms

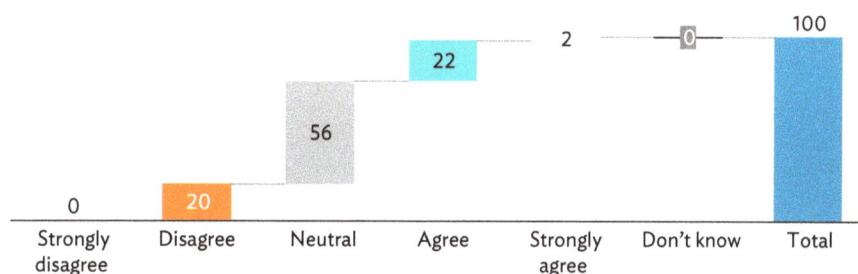

Common reasons for accelerated adoption:

Employees incentivized to upskill through COVID-related retraining schemes

Lack of labor due to movement restrictions necessitates more automation and shifting of activities to digital means

COVID-19 = coronavirus disease.
Note: Based on survey of employers in the transportation and storage industry between June and September 2021 (n=50).
Source: Asian Development Bank (Sustainable Development and Climate Change Department).

Skills Demand Analysis

Job Implications

The full adoption of 4IR technologies by firms in Azerbaijan is expected to create over 27,000 new jobs or the equivalent of 13% of the 2020 transportation and storage workforce, over and beyond the BAU growth of the industry's labor force, by 2025. This is as the productivity effect again outweighs the displacement effect. Job displacement due to automation is expected to displace around 66,500 positions but create around 94,000 new jobs, over and beyond BAU growth of the industry's labor force (Figure 21). As in the agro-processing industry, more transportation and storage jobs will be created in Azerbaijan by 2025 even without 4IR. The transportation and storage workforce grew at approximately 2.1% per annum[11] from 2015 and 2020, and if this growth were extrapolated up to 2025, an additional 23,000 jobs could have been created even without adopting 4IR technologies. However, the full adoption of 4IR technologies could see a total of around 50,000 more transportation and storage jobs in 2025 than in 2020. On the other hand, Box 2 shows the potential cost of government inaction.

Figure 21: Estimated Impact of Industry 4.0 on Number of Jobs by 2025 in the Transportation and Storage Industry in Azerbaijan

The adoption of 4IR technologies could lead to 13% more jobs in 5 years' time as the number of newly created jobs outweigh displaced jobs

Percent of jobs impacted due to displacement and productivity effects of 4IR in 5 years' time

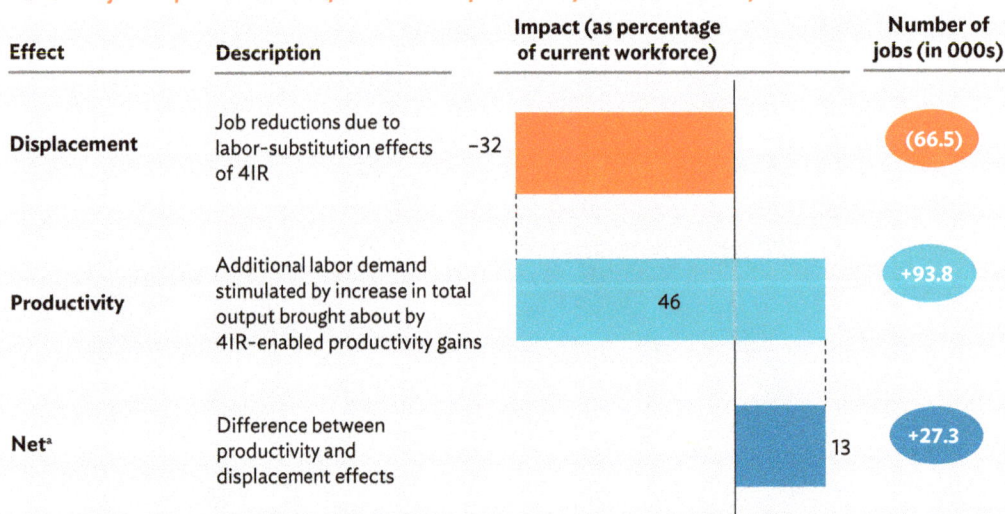

Effect	Description	Impact (as percentage of current workforce)	Number of jobs (in 000s)
Displacement	Job reductions due to labor-substitution effects of 4IR	−32	(66.5)
Productivity	Additional labor demand stimulated by increase in total output brought about by 4IR-enabled productivity gains	46	+93.8
Net[a]	Difference between productivity and displacement effects	13	+27.3

( ) = negative, 4IR = Fourth Industrial Revolution.

Note: Based on survey of employers in the transportation and storage industry between June and September 2021 (n=50).

Sources: Industry employment data from State Statistical Committee of Azerbaijan. Labour Market. https://www.stat.gov.az/source/labour/?lang=en; industry output data from State Statistical Committee of Azerbaijan. System of National Accounts and Balance of Payments. https://www.stat.gov.az/source/system_nat_accounts/?lang=en.

---

[11]   Calculated using industry and workforce data from State Statistical Committee of Azerbaijan.

Box 2: Potential Cost of Government Inaction on Job Creation under Industry 4.0 Adoption

The net gains estimated in this report assumes that the entire industry adopts Fourth Industrial Revolution (4IR) technologies. For this to occur, the Government of Azerbaijan has a crucial role to play in ensuring that appropriate policies are in place to encourage participation, especially for firms that face significant adoption barriers. Effective policies are also needed to smooth labor market frictions, where public reskilling initiatives could be necessary for displaced workers to successfully transition into newly created jobs. An interesting analysis may be to understand how the breadth and effectiveness of policies impact net job creation under 4IR technology adoption. This will require empirical evaluations (e.g., simulations of impact on labor productivity growth under different policy scenarios), which are beyond the scope of this report and could be the subject of future research.

A simple thought exercise to provide some immediate guidance could be to assume that only the proportion of firms (14% in the agro-processing industry and 20% in the transportation and storage industry), which indicated an intermediate or advanced understanding of 4IR technologies and their applications in 2020, would reap the productivity gains of 4IR. In this case, only 14% of the 15,000 net job gains (around 2,100 jobs) expected in the full adoption scenario will be realized for the agro-processing industry. Similarly, in the transportation and storage industry, only 20% of the 27,000 jobs (around 5,400 jobs) expected to be created in the full adoption scenario will be realized.

This research further demonstrates that the impact of 4IR goes beyond job creation. 4IR will change the nature of jobs in the agro-processing and transportation and storage industries and the types of skills that will be needed to take on these jobs. Past studies have demonstrated that this will free up time for workers doing routine tasks to take on higher-value tasks such as creative work. This would increase job satisfaction among workers and potentially lead to higher wages.

Source: AlphaBeta. 2017. *The Automation Advantage.* https://alphabeta.com/wp-content/uploads/2017/08/The-Automation-Advantage.pdf.

As in the agro-processing industry, the impact will differ by occupation. This country report categorized jobs in the transportation and storage industry into five occupational groups (see Table 5).

Table 5: Occupational Groups in the Information Technology–Business Process Outsourcing Industry

Occupational Group		Examples of Job Titles
1	**Technical**	• Technician • Website Designer
2	**Managerial**	• Chief Executive Officer • Warehouse Manager
3	**Customer-facing**	• Hotline Operator • Customer Service Executive
4	**Administrative**	• Secretary • Finance Executive
5	**Elementary and/or manual jobs**	• Warehouse Worker • Driver • Security Guard

Sources: Asian Development Bank (Sustainable Development and Climate Change Department) and AlphaBeta.

As in the agro-processing industry, most employers expect automation to decrease the number of jobs across all occupational groups, with many employers especially expecting decreases in manual and administrative job roles (Figure 22). Among the various occupational groups, the technical occupational group is where employers expect to see the highest increases in adoption of 4IR technologies, at 32%. The expected decrease in manual and administrative roles is consistent with the expectations that more IOT and systems integration technologies will be adopted. The number of workers to take on elementary or manual roles, such as security guards and warehouse workers, will be reduced as these processes will be automated via IOT sensors. Systems integration technologies will also reduce the amount of paperwork to be processed by shipping and logistics firms, and therefore the number of administrative staff. On the other hand, new technical roles could be created as more workers are needed to operate and maintain digital platforms or data collection systems. Technical workers are expected to replace manual workers as the largest occupational group by 2025 in the transportation and storage industry if 4IR technologies are fully adopted (Figure 23).

As in the agro-processing industry, male workers will benefit from a larger proportion of the job gains, compared to female workers with 1.8 times more of the new jobs created by 4R going to male workers (Figure 24). This is due to the relatively higher proportion of male workers in technical and managerial roles in which most employers expect to see new jobs created. According to the United Nations Development Programme (UNDP), past research reveals that cultural stereotypes continue to pose a barrier for women pursuing technical careers in Azerbaijan (UNDP 2018). While nearly half of all students enrolled in higher education institutions in Azerbaijan in 2020 were female, only 25% of students enrolled in technical and technological disciplines are female. Instead, most female students are enrolled in education-related disciplines.[12] Targeted policies to encourage and enable women to take up more technical and managerial roles are needed to ensure that female workers also reap the gains of 4IR adoption.

Figure 22: Expected Impact of Industry 4.0 on the Number of Jobs by 2025 in the Transportation and Storage Industry in Azerbaijan (%)

In the transportation and storage industry, most employers expect the number of technical jobs to increase with the adoption of 4IR technologies

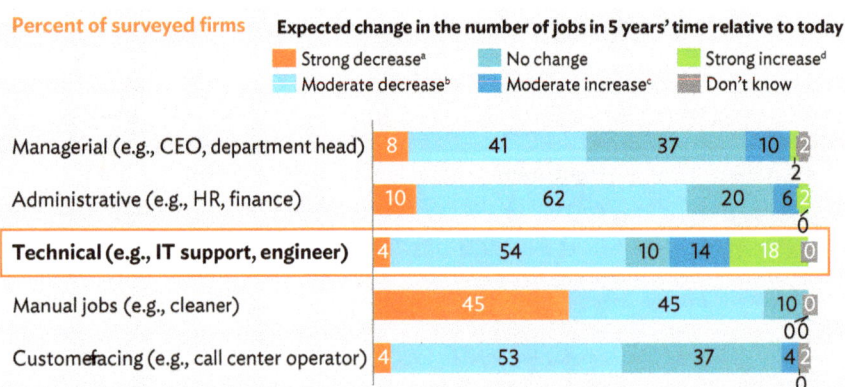

Percent of surveyed firms

Expected change in the number of jobs in 5 years' time relative to today

- Strong decrease[a]
- Moderate decrease[b]
- No change
- Moderate increase[c]
- Strong increase[d]
- Don't know

Occupational group	Strong decrease	Moderate decrease	No change	Moderate increase	Strong increase	Don't know
Managerial (e.g., CEO, department head)	8	41	37	10	2	2
Administrative (e.g., HR, finance)	10	62	20	6	2	0
Technical (e.g., IT support, engineer)	4	54	10	14	18	0
Manual jobs (e.g., cleaner)	45	45	10	0	0	0
Customer-facing (e.g., call center operator)	4	53	37	4	2	0

CEO = chief executive officer, HR = human resources, IT = information technology.

Notes: Based on survey of employers in the transportation and storage industry between June and September 2021 (n=50).

a   Strong decrease: Greater than or equal to 50% decrease in number of jobs.
b   Moderate decrease: Less than 50% decrease in number of jobs.
c   Moderate increase: Less than 50% increase in number of jobs.
d   Strong increase: Greater than or equal to 50% increase in number of jobs.

Source: Asian Development Bank (Sustainable Development and Climate Change Department).

---

[12]   State Statistical Committee of Azerbaijan. Education Indicators. https://www.stat.gov.az/source/education/?lang=en.

Figure 23: Composition of Jobs in 2020 and by 2025, by Occupational Group in the Transportation and Storage Industry in Azerbaijan (%)

The distribution of jobs will change, with technical jobs seeing the largest increase in 5 years' time

Weighted average percentage share of employees by occupational group in surveyed firms

● Negative shift
● Positive shift

Occupational group	Share today	Share in 5 years' timeª	Percentage shift
Manual jobs	35	28	(6.7%)
Administrative	12	12	0
Customer-facing	14	15	+0.6%
Managerial	12	13	+1.0%
Technical	28	33	+5.1%

Notes: Based on survey of employers in the transportation and storage industry between June and September 2021 (n=50). Average share of employees in surveyed firms is weighted by the number of employees in each firm, as indicated by respondents; percentages might not add up to 100% due to rounding. The change in the number of workers in each job type is based on the number of firms indicating different levels of changes in number of jobs, i.e., "strong increase," "moderate increase," "no change," "moderate decrease," "strong decrease." The midpoint of the range for each option for expected change is used; for expected increase or decrease of over 50%, the lower bound of 50% is used.
Source: Asian Development Bank (Sustainable Development and Climate Change Department).

Figure 24: Estimated Net Job Gains by Gender from Industry 4.0 Adoption by 2025 in the Transportation and Storage Industry in Azerbaijan

Policies to increase female participation in the transportation and storage sector are critical to ensure that women also benefit from job gains from 4IR adoption

Estimated number of net jobs created by gender (in thousands)

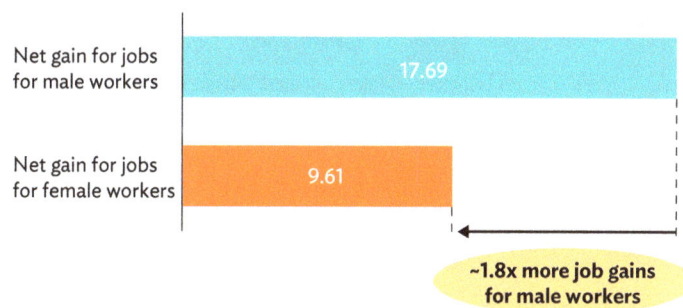

Net gain for jobs for male workers	17.69
Net gain for jobs for female workers	9.61

~1.8x more job gains for male workers

4IR = Fourth Industrial Revolution.
Note: Based on survey of employers in the transportation and storage industry between June and September 2021 (n=50).
Source: Industry employment data from State Statistical Committee of Azerbaijan. Labour Market. https://www.stat.gov.az/source/labour/?lang=en.

Apart from changing the types of jobs that are available, the adoption of 4IR technologies could also make existing jobs safer. One example is the Daimler's Highway Pilot System that increases road safety by relieving the driver during potentially dangerous sections of a route. Daimler's research found that assistance from onboard technology could reduce driver sleepiness by 25% compared with driving a conventional truck and reduce accidents caused by human error (AIG 2017).

Task Implications

The employer survey reveals a sharp drop in the proportion of time that employers expect workers to spend on routine tasks, both physical and interpersonal, by 2025 with the adoption of technologies (Figure 25). This is consistent with the change in the distribution of jobs across various occupational groups. Manual and administrative roles such as stock-taking in the warehouse, or processing paperwork such as payment invoices, tend to consist of routine tasks, which are expected to decrease in proportion. Instead, the largest job gains are expected in technical roles that likely require using digital platforms and analytical skills.

Figure 25: Time Spent by Employees on Tasks at Work in 2020 and by 2025 in the Transportation and Storage Industry in Azerbaijan (%)

Adoption of 4IR technologies is expected to shift the distribution of weekly working hours from routine to analytical and nonroutine tasks

Average percentage share of weekly working hours spent by task in surveyed firms

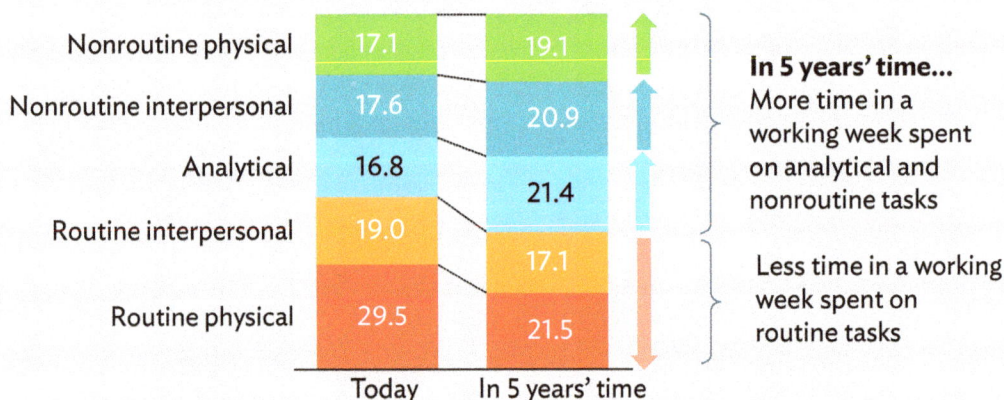

	Today	In 5 years' time
Nonroutine physical	17.1	19.1
Nonroutine interpersonal	17.6	20.9
Analytical	16.8	21.4
Routine interpersonal	19.0	17.1
Routine physical	29.5	21.5

In 5 years' time... More time in a working week spent on analytical and nonroutine tasks

Less time in a working week spent on routine tasks

4IR = Fourth Industrial Revolution.
Note: Based on survey of employers in the transportation and storage industry between June and September 2021 (n=50). Figures are average percentage shares of weekly working hours spent on tasks, and include rounding adjustments.
Source: Asian Development Bank (Sustainable Development and Climate Change Department).

Skills Implications

The shift toward more analytical tasks is also reflected in the changes in the skills sought by employers. In 2025, digital and ICT skills, creative thinking and/or design, as well as numeracy skills will be most sought after by employers (Figure 26). The ability of 4IR technologies, particularly systems integration and blockchain technologies, to reduce human interaction in the transportation and storage industry is also reflected in the findings. Skills such as written and verbal communication and social and interpersonal skills will drop significantly in relative importance. Digital twin systems used in warehouses and port facilities to consolidate comprehensive data on the movement of inventory, equipment, and personnel can aid the efficiency of operations and reduce the need for constant reporting and human interaction (footnote 13). Comparing the skills that are prioritized by employers for 4IR adoption by 2025 against skills in which a step-up in proficiency is needed, the analysis shows that workers would require a significant step-up in skills relevant to 4IR adoption such as numeracy and digital and/or ICT skills from 2020 (Figure 27). Among the employers that feel that a step-up from basic proficiency in digital and/or ICT skills is needed, 24% feel that a step-up to the intermediate level of proficiency is needed while 76% would like to see a step-up to advanced proficiency. A step-up to an advanced level of proficiency would be significantly more difficult to achieve and could also hint at industry-specific digital skills (e.g., the ability to use supply chain tracking software). As such, policy makers would need to work closely with training institutions and employers to ensure that sufficient training is provided to workers. In contrast, while creative thinking and/or design skills will be increasingly valued by employers by 2025, most employers assessed the current proficiency levels of employees in this area to be sufficient. This could be due to the relatively limited demand for creative workers in the industry in 2020 based on the current level of 4IR adoption, so that employers are unable to visualize a future need for such workers at present.

Figure 26: Importance of Skills in 2020 and for Industry 4.0 Adoption by 2025 in the Transportation and Storage Industry in Azerbaijan

Digital and ICT skills will be the most important skill for 4IR technology adoption in the industry in 5 years' time

- Skills of increasing importance in 5 years' time
- Skills of decreasing importance in 5 years' time
- Skills with no change in importance in 5 years' time

Importance ranking	Today[a]	In 5 years' time[b]	Change in ranking
1	Written communication	Digital and/or ICT skills	+1
2	Digital and/or ICT skills	Creative thinking or design	+7
3	Critical thinking	Numeracy	+5
4	Social and interpersonal	Complex problem solving	+3
5	Verbal communication	Adaptive learning	+5
6	Management	Critical thinking	(3)
7	Complex problem solving	Management	(1)
8	Numeracy	Social and interpersonal	(4)
9	Creative thinking or design	Verbal communication	(4)
10	Adaptive learning	Written communication	(9)

( ) = negative, 4IR = Fourth Industrial Revolution, ICT = information and communication technology.
Note: Based on survey of employers in the transportation and storage industry between June and September 2021 (n=50); job data on the transportation and storage industry from the job portal Offer.AZ (accessed June 2021).
[a] Evaluated using the employer survey and supported by job portal data.
[b] Evaluated using the employer survey.
Source: Asian Development Bank (Sustainable Development and Climate Change Department).

Figure 27: Required Step-Up in Employee Proficiency Level from 2020 for Industry 4.0 Adoption by 2025 in the Transportation and Storage Industry in Azerbaijan (%)

To be 4IR-ready, workers would require proficiency leaps in numeracy skills

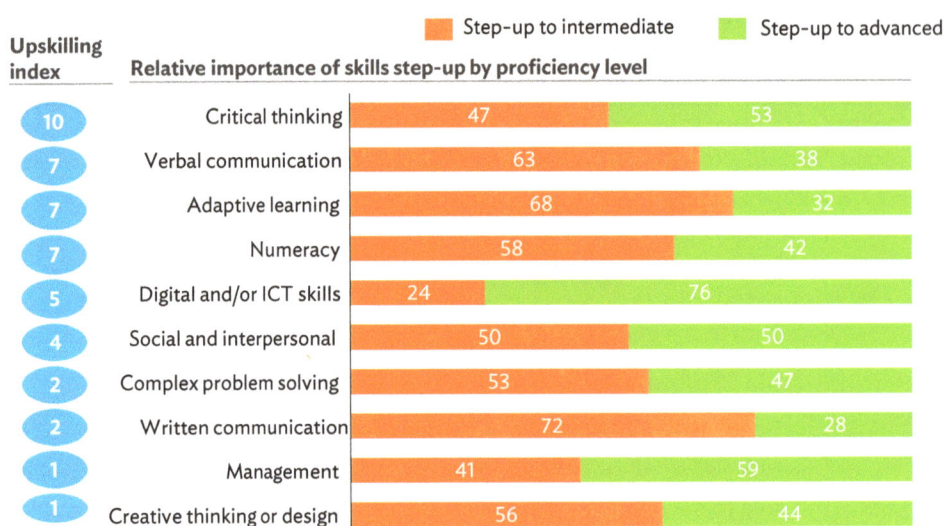

| | Step-up to intermediate | Step-up to advanced |

Upskilling index	Relative importance of skills step-up by proficiency level		
10	Critical thinking	47	53
7	Verbal communication	63	38
7	Adaptive learning	68	32
7	Numeracy	58	42
5	Digital and/or ICT skills	24	76
4	Social and interpersonal	50	50
2	Complex problem solving	53	47
2	Written communication	72	28
1	Management	41	59
1	Creative thinking or design	56	44

4IR = Fourth Industrial Revolution, ICT = information and communication technology.
Note: Based on survey of employers in the transportation and storage industry between June and September 2021 (n=50).
Source: Asian Development Bank (Sustainable Development and Climate Change Department).

Skills Supply Trends

Like the agro-processing industry, employers in the transportation and storage industry find it challenging to identify and recruit good quality candidates for job openings (Figure 28). A majority also agree that there is significant variance in the quality of graduates from different training institutions. Despite the difficulties in recruiting good quality candidates, it appears that most employers are either unwilling to provide training or lack the resources to do so. Only around of half of employers agree that employees receive sufficient training to do their jobs well, and only around a third of firms surveyed said they currently invest sufficiently in training their workers (Figure 29). Around 44% of employers indicated that it was difficult to find good-quality training providers. During the country consultations conducted, industry stakeholders said that many firms in Azerbaijan were relatively small and lacked the resources to invest substantially in worker training. Investments in skills development often took a long time to reap gains and small companies at an early stage might not see the long-term gains of such investments. Over half of all enterprises in Azerbaijan are small enterprises with fewer than 25 employees (Aliyev 2019). This suggests that firms in the transportation and storage could require financial support or incentives to train their workers, particularly as skill needs change rapidly with 4IR technologies being adopted.

On-the-job training is currently the most common training channel for firms in the transportation and storage industry. However, the use of professional and formal education courses is likely to increase by 2025, suggesting that firms see a need to provide more formal training to workers—both for short-term courses lasting less than 6 months and for part-time degrees or diplomas—to meet the changing demand in skills created by the adoption of 4IR technologies. Employers also expect online training to fall slightly, as employers expect some online training courses to return to in-person training as the COVID-19 situation improves.

Figure 28: Employer Sentiment Toward Graduates Hired in the Transportation and Storage Industry in Azerbaijan (%)

About 58% of employers disagree that it is easy to identify and recruit high quality graduates for entry-level positions

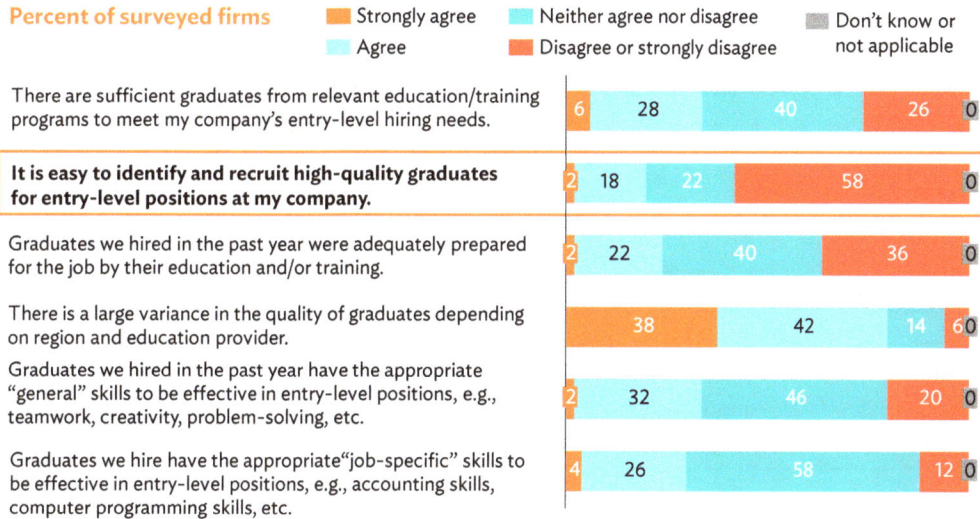

Percent of surveyed firms — Strongly agree | Neither agree nor disagree | Don't know or not applicable | Agree | Disagree or strongly disagree

Statement	Strongly agree	Agree	Neither	Disagree or strongly disagree	Don't know
There are sufficient graduates from relevant education/training programs to meet my company's entry-level hiring needs.	6	28	40	26	0
It is easy to identify and recruit high-quality graduates for entry-level positions at my company.	2	18	22	58	0
Graduates we hired in the past year were adequately prepared for the job by their education and/or training.	2	22	40	36	0
There is a large variance in the quality of graduates depending on region and education provider.	38	42	14	6	0
Graduates we hired in the past year have the appropriate "general" skills to be effective in entry-level positions, e.g., teamwork, creativity, problem-solving, etc.	2	32	46	20	0
Graduates we hire have the appropriate "job-specific" skills to be effective in entry-level positions, e.g., accounting skills, computer programming skills, etc.	4	26	58	12	0

Note: Based on survey of employers in the transportation and storage industry between June and September 2021 (n=50).
Source: Asian Development Bank (Sustainable Development and Climate Change Department).

Figure 29: Employers' Perception on Training for Employees in the Transportation and Storage Industry in Azerbaijan (%)

Close to half of employers find it difficult to find good quality training providers to train their employees

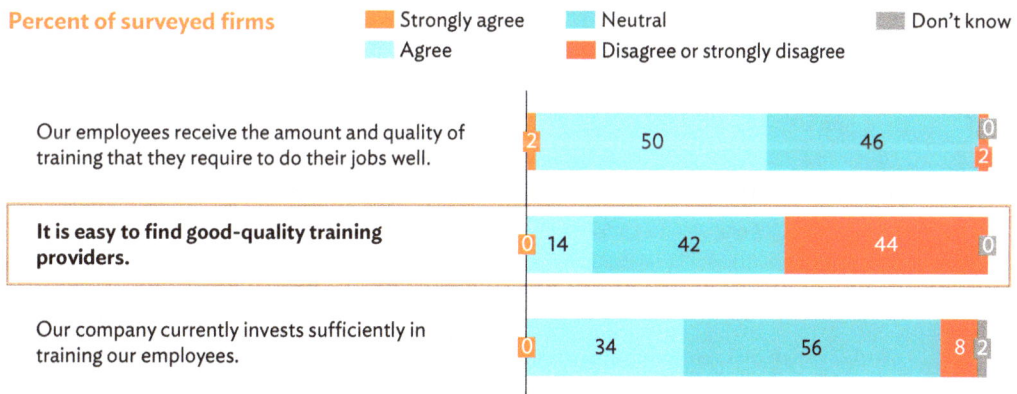

Percent of surveyed firms — Strongly agree | Neutral | Don't know | Agree | Disagree or strongly disagree

Statement	Strongly agree	Agree	Neutral	Disagree or strongly disagree	Don't know
Our employees receive the amount and quality of training that they require to do their jobs well.	2	50	46	0	2
It is easy to find good-quality training providers.	0	14	42	44	0
Our company currently invests sufficiently in training our employees.	0	34	56	8	2

Note: Based on survey of employers in the transportation and storage industry between June and September 2021 (n=50
Source: Asian Development Bank (Sustainable Development and Climate Change Department).

Figure 30: Proportion of Employees Receiving Training in 2020 and Requiring Training by 2025 in the Transportation and Storage Industry in Azerbaijan (%)

The proportion of employees undergoing professional and formal education courses is expected to increase in five years' time

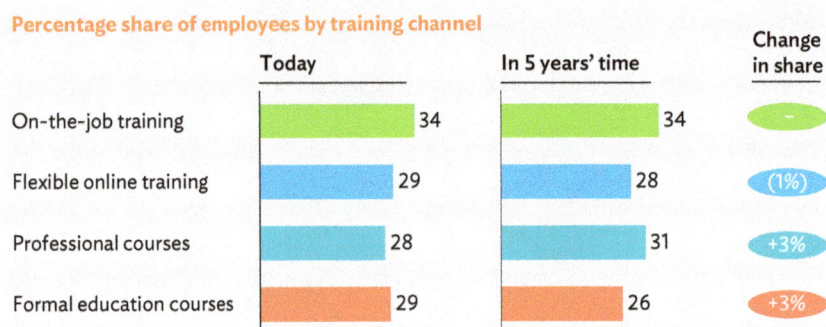

Percentage share of employees by training channel

	Today	In 5 years' time	Change in share
On-the-job training	34	34	–
Flexible online training	29	28	(1%)
Professional courses	28	31	+3%
Formal education courses	29	26	+3%

( ) = negative.
Note: Based on survey of employers in the transportation and storage industry between June and September 2021 (n=50). The sum of all shares for 2020 and by 2025 exceeds 100%, as one employee can undergo training in more than one training channel.
Source: Asian Development Bank (Sustainable Development and Climate Change Department).

E.   Emerging Jobs

Through the employer surveys conducted and scrapping of online job portals, it emerged that employers expect a variety of new job roles to become more prevalent as 4IR technologies are increasingly adopted across business functions. In the agro-processing industry, these include digital marketing managers, Big Data specialists, and machine learning experts. In the transportation and storage industry, jobs related to AI and digital marketing are expected to be created (Figure 31).

Figure 31: Job Roles Expected to Become More Prominent with the Adoption of Industry 4.0 Technologies by 2025

The adoption of 4IR technologies will lead to different job roles in both industries according to employer surveys and online job portal analysis

Agro-processing

Digital marketing or e-commerce managers: engage clients through online channels to market processed food products; design and maintain website or other e-commerce channels

Machine learning experts: develop artificial intelligence based algorithms and devices that enable machine learning (e.g., Artificial Intelligence-enabled systems to check food quality)

Big Data specialists: set up infrastructure for data collection and analysis, integrate data from various resources, and analyze data to predict consumer trends

Transportation and storage

Artificial intelligence engineers: develop artificial intelligence based algorithms for route planning and optimization

Digital marketing managers: engage clients through online channels to market transportation and logistics services

4IR = Fourth Industrial Revolution.
Note: Based on survey of employers conducted in Azerbaijan between June and September 2021 (n=50 for agro-processing, n=50 for transportation and storage); jobs data from the job portal Offer.AZ (accessed June 2021).
Source: Asian Development Bank (Sustainable Development and Climate Change Department).

2 Overview of the Training Landscape

This chapter provides insights into the performance of the technical and vocational education and training (TVET) sector in Azerbaijan as it prepares to deal with the challenges emerging from 4IR technology adoption. The insights are drawn from a survey of training institutions in Azerbaijan, complemented with insights from the employer surveys discussed in Chapter 1.

Overall, training institutions in Azerbaijan are not well prepared for 4IR and would require significant support from the government. Around 54% of training institutions surveyed strongly agree that technical and financial support is needed to enable them to prepare workers for 4IR. As of 2020, only a small proportion of training institutions teach courses specific to 4IR technologies or use 4IR technologies to deliver training. Training institutions would require support to build up the capabilities of their teaching staff to ensure that they are equipped to teach 4IR-related courses as well as implement 4IR-enabled teaching approaches in the classroom.

The training landscape analysis also revealed the need for stronger alignment on industry's skills needs between employers and training institutions. While 83% of training institutions assess their graduates to be adequately prepared for entry-level positions, less than a quarter of employers in the agro-processing and transportation and storage industries take the same view. Stronger collaboration in the design and implementation of training curricula and programs will be particularly critical in Azerbaijan. Currently, only 20% of training institutions gather input from industry stakeholders to design training curricula.

In summary, significant policy action is needed to ready training institutions to prepare workers for 4IR and ensure that their graduates can gain access to quality jobs. Policies focused on building strong alignment between employers and training institutions on future skill needs will be particularly critical as the adoption of 4IR technologies could lead to rapid changes in the skills that employers require from workers.

To better understand the supply of talent and skills for the adoption of 4IR technology, a survey of 70 training institutions was undertaken in Azerbaijan. These included public and private institutions of higher learning as well as TVET institutions. Of the training institutions surveyed, 97% trained at least 100 students per year.

A.   Industry 4.0 Readiness and the Impact of COVID-19

In contrast to employers, training institutions demonstrated slightly more confidence in their preparedness for 4IR. Only 14% of firms in the agro-processing industry and 20% of firms in the transportation and storage industry reported a good understanding of 4IR. However, 10% of training institutions surveyed strongly agree and 47% agree that they have a good understanding of the skills that will be needed to be developed to prepare graduates for 4IR (Figure 32). Most training institutions also believe that they can adequately prepare workers for 4IR as per ongoing plans but would need additional technical and financial support. This could be due to training institutions being unclear on the types of 4IR technologies and industry-specific applications prioritized by firms in Azerbaijan, or being unable to engage trainers who have the relevant knowledge. In addition, training institutions could also lack the financial resources to purchase software or equipment to train students in 4IR technologies.

Figure 32: Perception of Training Institutions on Readiness for Industry 4.0 in Azerbaijan (%)

About 54% of training institutions strongly agree that they need technical and financial support to prepare workers for 4IR

Percent of surveyed training institutions

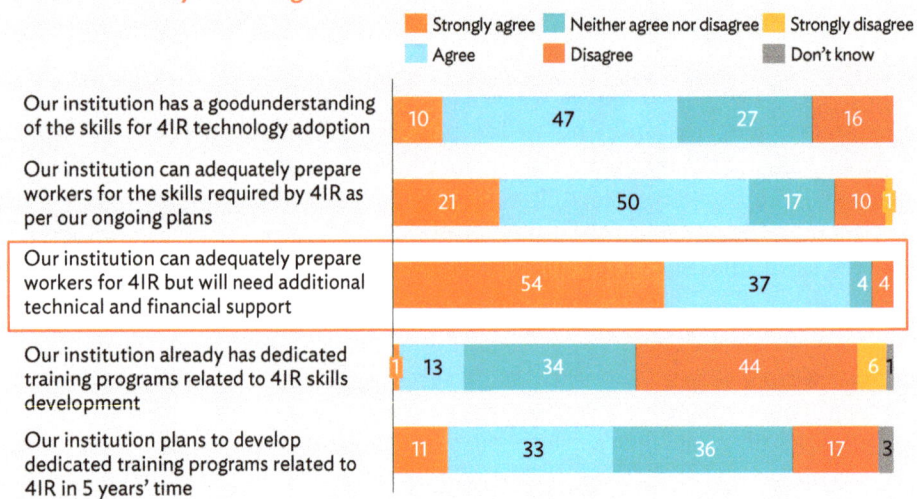

Legend: Strongly agree · Neither agree nor disagree · Strongly disagree · Agree · Disagree · Don't know

Statement	Strongly agree	Agree	Neither agree nor disagree	Disagree	Strongly disagree	Don't know
Our institution has a goodunderstanding of the skills for 4IR technology adoption	10	47	27	16		
Our institution can adequately prepare workers for the skills required by 4IR as per our ongoing plans	21	50	17	10	1	
Our institution can adequately prepare workers for 4IR but will need additional technical and financial support	54	37	4	4		
Our institution already has dedicated training programs related to 4IR skills development	1	13	34	44	6	1
Our institution plans to develop dedicated training programs related to 4IR in 5 years' time	11	33	36	17		3

4IR = Fourth Industrial Revolution.
Note: Based on survey of training institutions between June and September 2021 (n = 70).
Source: Asian Development Bank (Sustainable Development and Climate Change Department).

Most training institutions in Azerbaijan indicated that they were able to continue training activities during the COVID-19 pandemic. Only 9% of training institutions surveyed said they had to fully close for some time due to the inability to conduct in-person training, while 99% managed to shift some or most of their courses online (Figure 33).

Figure 33: Impact of COVID-19 on Training Institutions in Azerbaijan (%)

Close to 100% of training institutions indicated that they shifted some or most of their courses online in response to COVID-19

Percent of surveyed training institutions

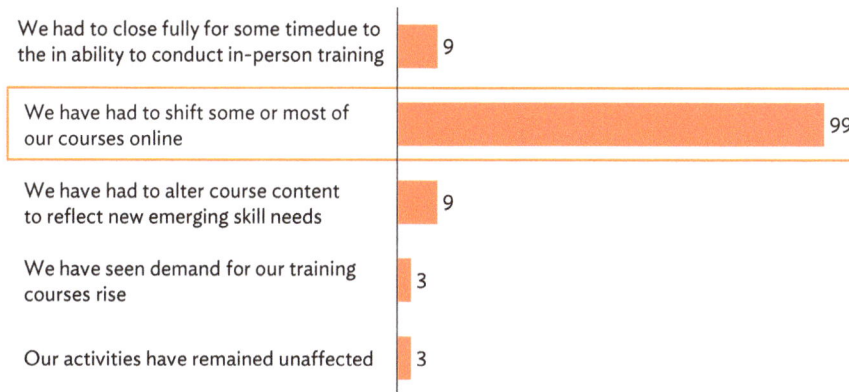

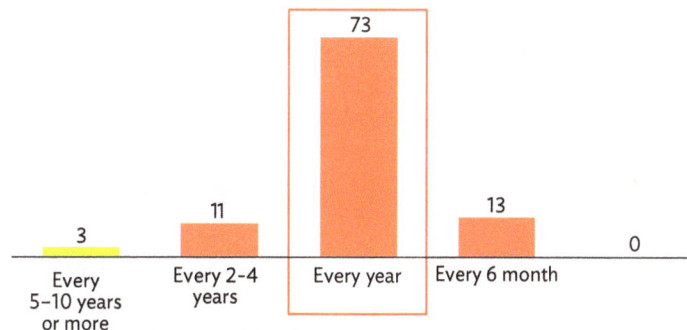

We had to close fully for some timedue to the in ability to conduct in-person training	9
We have had to shift some or most of our courses online	99
We have had to alter course content to reflect new emerging skill needs	9
We have seen demand for our training courses rise	3
Our activities have remained unaffected	3

COVID-19 = coronavirus disease.
Note: Based on survey of training institutions between June and September 2021 (n = 70).
Source: Asian Development Bank (Sustainable Development and Climate Change Department).

B. Curricula

Industry 4.0 technologies can transform the workplace and jobs quickly. Hence, it is critical that training curricula be updated frequently to meet employers' needs. In Azerbaijan, 73% of training institutions claim that they review and update their training curricula annually (Figure 34).

Figure 34: Frequency of Review and Update of Curricula by Training Institutions in Azerbaijan (%)

About 73% of training institutions review and update their curricula annually

Percent of surveyed training institutions

- Every 5–10 years or more: 3
- Every 2–4 years: 11
- Every year: 73
- Every 6 month: 13
- 0

Note: Based on survey of training institutions between June and September 2021 (n=70).
Source: Asian Development Bank (Sustainable Development and Climate Change Department).

Despite the frequent updates made to the training curricula, only 13% of the training institutions surveyed currently teach courses related to 4IR, and only 29% provide general digital skills programs to improve digital literacy (Figure 35). This could be in part due to the limited demand for 4IR-related courses given the limited adoption of 4IR technologies among firms in Azerbaijan, as illustrated by the surveys conducted in the agro-processing and transportation and storage industries. However, during the country consultations, local experts and stakeholders highlighted the potential demand for ICT and 4IR-related courses given the current lack of such courses in Azerbaijan, stressing that it would be particularly important for public training providers to provide relevant courses to ensure that they are accessible to a large pool of students. Encouragingly, close to half of training institutions use online self-learning modules although only a small proportion use simulators or systems based on augmented reality (AR)/virtual reality (VR) technologies to deliver training.

Of the firms that provide courses specific to 4IR technologies, most appear to offer courses in systems integration technologies that link together different computing systems and software applications (Figure 36). Some 46% of firms surveyed in the transportation and storage industry expect to adopt systems integration technologies to link various systems including sales, warehousing, freight, as well last-mile fulfillment systems during 2020–2025. Training institutions could strengthen their focus in this area. Another 4IR technology that training institutions could focus on is IOT technologies. Over 50% of employers in the agro-processing and transportation and storage industries intend to adopt IOT technologies, but only 4% of training institutions offer relevant courses currently. To ensure that training institutions can provide courses aligned to changing skills needs created by 4IR, Azerbaijan could consider the development of 4IR action plans that provide information on technology impacts, labor market shifts, the skills required for different occupations, and reskilling options for different industries. These action plans could build on the existing strategic road maps to focus on how the adoption of 4IR technologies can help to achieve long-term growth objectives and strengthen alignment on future plans between employers, training institutions, and policy makers.

Figure 35: Prevalence of Industry 4.0-Related Courses and Industry 4.0-Based Delivery in Training Institutions in Azerbaijan (%)

Currently, only a small proportion of training institutions teach courses specific to 4IR technologies and/or use 4IR technologies to deliver training

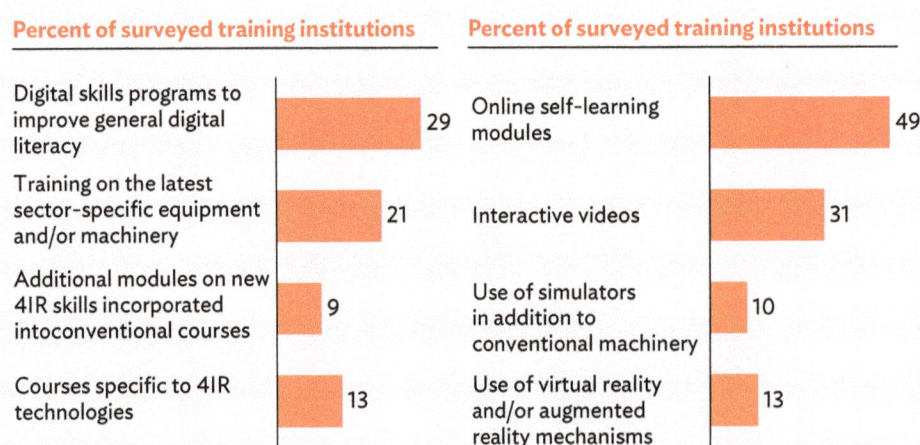

Percent of surveyed training institutions

Digital skills programs to improve general digital literacy	29
Training on the latest sector-specific equipment and/or machinery	21
Additional modules on new 4IR skills incorporated intoconventional courses	9
Courses specific to 4IR technologies	13

Percent of surveyed training institutions

Online self-learning modules	49
Interactive videos	31
Use of simulators in addition to conventional machinery	10
Use of virtual reality and/or augmented reality mechanisms	13

4IR = Fourth Industrial Revolution.
Note: Based on survey of training institutions between June and September 2021 (n=70). Percentages do not add up to 100% as respondents were asked to select all options that apply.
Source: Asian Development Bank (Sustainable Development and Climate Change Department).

Figure 36: Planned Adoption of Specific Industry 4.0 Technologies by Employers and Prevalence of Courses Relevant to These Technologies in Training Institutions in Azerbaijan

Over half of employers in both industries plan to adopt IOT technologies but only 4% of training institutions offer relevant courses

Percent of employers that plan to adopt the technology in 5 years' time;[a]
Percent of surveyed training institutions[b]

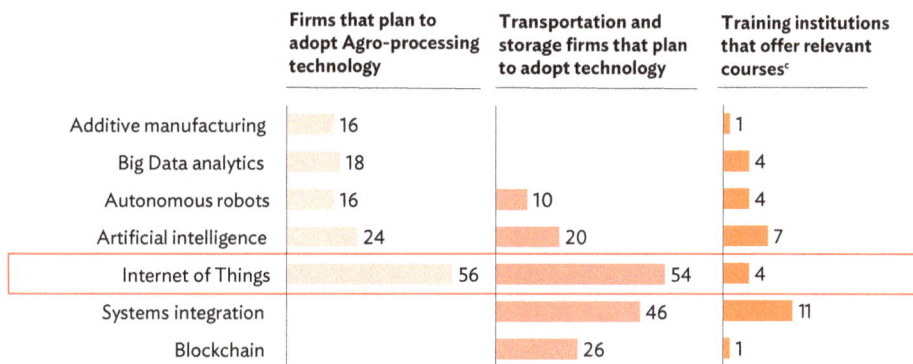

	Firms that plan to adopt Agro-processing technology	Transportation and storage firms that plan to adopt technology	Training institutions that offer relevant courses[c]
Additive manufacturing	16		1
Big Data analytics	18		4
Autonomous robots	16	10	4
Artificial intelligence	24	20	7
Internet of Things	56	54	4
Systems integration		46	11
Blockchain		26	1

[a] Based on survey of employers in the agro-processing industry between June and September 2021 (n=50).
[b] Based on survey of employers in the transportation and storage industry between June and September 2021 (n=50).
[c] Based on survey of training institutions between June and September 2021 (n=70).
Notes: Values are percentages of firms that responded "moderate" or "high" to current deployment of the technology. Percentages do not add up to 100% as respondents were asked to select all options that apply.
Source: Asian Development Bank (Sustainable Development and Climate Change Department).

C.   Industry Engagement

Policy makers recognize the importance of stronger collaboration between industry and training institutions, and one of the key targets of the Strategic Roadmap for Vocational Education and Training adopted by the Government of Azerbaijan in 2016 is the integration of employers into the vocational education and training (VET) system. Sectoral committees for skills development have also been established to allow training institutions, employers, and policy makers to collaborate on developing occupational standards and qualifications. Despite these efforts, there appears to be limited partnership between employers and training institutions. Only 20% of training institutions surveyed gather input for curricula from industry stakeholders and only 26% offer teaching placements for industry professionals at training institutions (Figure 37). While the State Agency on Vocational Education has also initiated more than 100 cooperation agreements between the agency, schools, and the private sector to encourage more partnership activities between employers and training institutions over the last few years, these are largely broad agreements to cooperate and do not bind parties to specific actions (European Training Foundation 2020a). As such, actual partnership activities appear to be limited and only 19% of training institutions surveyed organize workplace-based training for students.

Figure 37: Partnership Activities between Training Institutions and Employers in Azerbaijan (%)

Only 4% of training institutions work with employers to organize job fairs to advertise job opportunities

Percent of training institutions

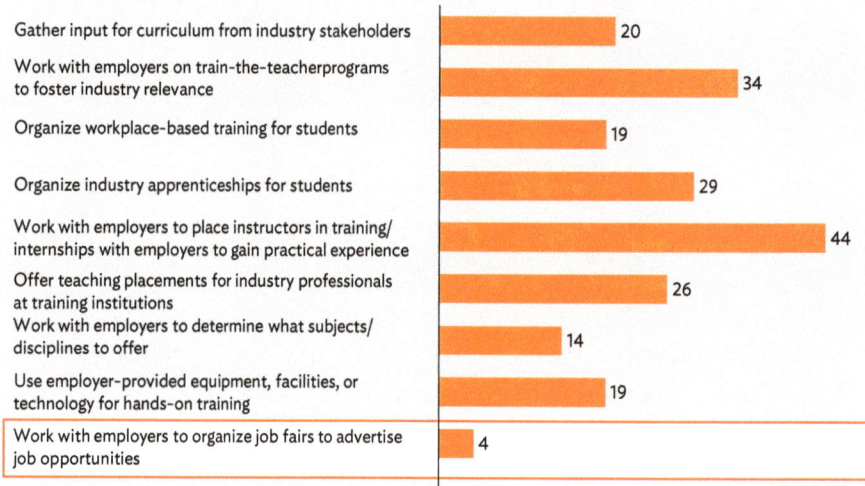

Gather input for curriculum from industry stakeholders	20
Work with employers on train-the-teacher programs to foster industry relevance	34
Organize workplace-based training for students	19
Organize industry apprenticeships for students	29
Work with employers to place instructors in training/internships with employers to gain practical experience	44
Offer teaching placements for industry professionals at training institutions	26
Work with employers to determine what subjects/disciplines to offer	14
Use employer-provided equipment, facilities, or technology for hands-on training	19
Work with employers to organize job fairs to advertise job opportunities	4

Note: Based on survey of training institutions between June and September 2021 (n = 70). Percentages do not add up to 100% as respondents were asked to select all options that apply.

Source: Asian Development Bank (Sustainable Development and Climate Change Department).

However, while training institutions report a low level of industry engagement overall, the employer surveys suggest that firms in the transportation and storage industry engage training institutions more actively than firms in the agro-processing industry. While only 20% of agro-processing firms surveyed provide input to incorporate the latest industry knowledge in training curricula, 40% of transportation and storage employers do the same (Figure 38). Interviews with industry experts similarly suggested that firms in the transportation and storage industry were keen to engage in dialogue with training institutions to address their skills needs; and that international development partners such as GIZ (Deutsche Gesellschaft für Internationale Zusammenarbeit) were providing support to develop curricula and training materials for selected occupations in the industry.[13] Encouragingly, despite the limited partnership activities between employers and training institutions, there appears to be frequent communication. Close to 90% of employers in both industries communicate with training institutions once a year or more frequently (Figure 39). These regular communication channels could be leveraged to drive partnership in more areas.

---

[13]   Expert interviews; GIZ. Private Sector Development and Technical Vocational Education and Training, South Caucasus. https://www.giz.de/en/worldwide/20324.html.

Figure 38: Partnership Activities Between Employers and Training Institutions in Azerbaijan (%)

Less than half of employers provide input to training institutions to incorporate the latest industry knowledge in their training curricula

Percent of surveyed firms

■ Yes ■ No, but willing to explore this

Activity	Agro-processing (Yes / No)	Transportation and storage (Yes / No)
Provide input to incorporate the latest industry knowledge in training curricula	20 / 78	40 / 58
Provide train-the-teacher programs to instructor to build relevant industry knowledge	46 / 40	34 / 54
Organize workplace-based training courses for students	36 / 36	36 / 48
Organize industry apprenticeships for students	38 / 54	54 / 40
Offer instructors in-house training or internships to gain hands-on experience	38 / 58	34 / 58
Incentivize staff to take up teaching part-time or go on teaching secondments	30 / 48	20 / 50
Work with training institutions to determine courses to offer	16 / 64	16 / 70
Provide equipment or facilities for institutions to provide students with hands-on training	38 / 44	42 / 54
Participate in or organize job fairs to advertise entry-level roles	44 / 38	52 / 32
Conducts active partnership activities with training institutions	40 / 48	50 / 48

Note: Based on survey of employers in the agro-processing industry (n = 50) and transportation and storage industry (n=50) between June and September 2021.
Source: Asian Development Bank (Sustainable Development and Climate Change Department).

Figure 39: Frequency of Communication between Employers and Training Institutions in Azerbaijan (%)

The majority of transportation and storage and agro-processing employers communicate with training institutions once a year or more

Percent of surveyed firms

■ Transportation and storage ■ Agro-processing

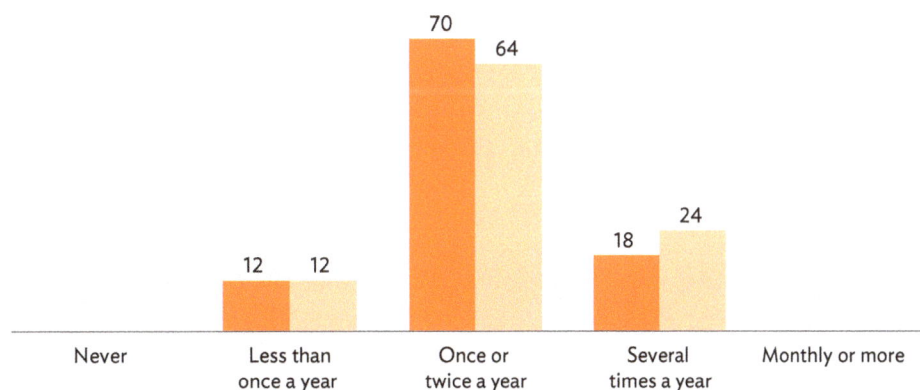

	Never	Less than once a year	Once or twice a year	Several times a year	Monthly or more
Transportation and storage		12	70	18	
Agro-processing		12	64	24	

Notes: Based on survey of employers in the agro-processing industry (n=50) and transportation and storage industry (n=50) between June and September 2021. Percentages may not add up to 100% due to rounding.
Source: Asian Development Bank (Sustainable Development and Climate Change Department).

D. Teachers, Trainers, and Instructors

The training institution survey revealed some gaps in the efforts of training institutions to ensure the quality of teaching staff. Only 34% of training institutions surveyed conduct reviews of instructors' performance annually or more frequently, and only 9% conduct frequent feedback sessions with instructors (Figure 40). These limited avenues of feedback means that training institutions would not be able to identify capability gaps of instructors and develop them professionally. Against the backdrop of the adoption of 4IR technology changing jobs and skills needs in the workforce, more professional development initiatives would also be needed to ensure that instructors can take on new training approaches and instruct students in new technologies. Currently, less than half of training institutions surveyed organize ongoing professional development and training for instructors.

Figure 40: Training Institutions' Practices to Support Instructors in Azerbaijan (%)

Only 9% of training institutions conduct frequent feedback sessions with instructors

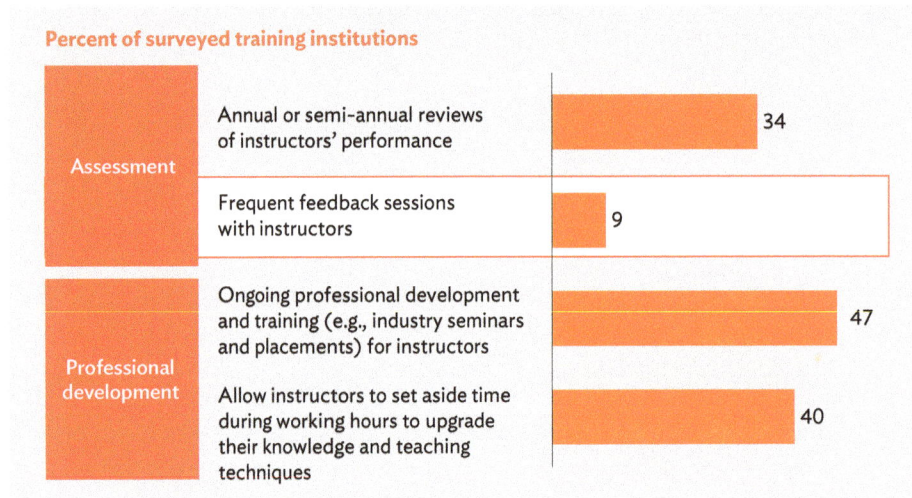

Percent of surveyed training institutions

Assessment	Annual or semi-annual reviews of instructors' performance	34
	Frequent feedback sessions with instructors	9
Professional development	Ongoing professional development and training (e.g., industry seminars and placements) for instructors	47
	Allow instructors to set aside time during working hours to upgrade their knowledge and teaching techniques	40

Note: Based on survey of training institutions between June and September 2021 (n=70). Percentages do not add up to 100% as respondents were asked to select all options that apply.
Source: Asian Development Bank (Sustainable Development and Climate Change Department).

E. Performance and Policy Support

Training institutions were asked to comment on their current performance and the types of policy support that would be required. About 30% of the training institutions found it at least somewhat difficult to fill training places, attributing this to trainees being unaware of courses available at their institutions (Figure 41). A clear skills development road map for each industry that sets out the skills needs for specific occupations and relevant training courses available could better match potential trainees and job seekers to training and jobs. In terms of impactful public policies, training institutions cited more government funding for more students to take up courses as well as support for online course delivery mechanisms as useful policy levers (Figure 42).

Figure 41: Training Institutions' Perceptions on and Reasons for Difficulty in Filling Places in Azerbaijan (%)

About 30% of training institutions find it difficult to fill places. The key reason is lack of awareness of programs offered by the institutions

Percent of surveyed training institutions

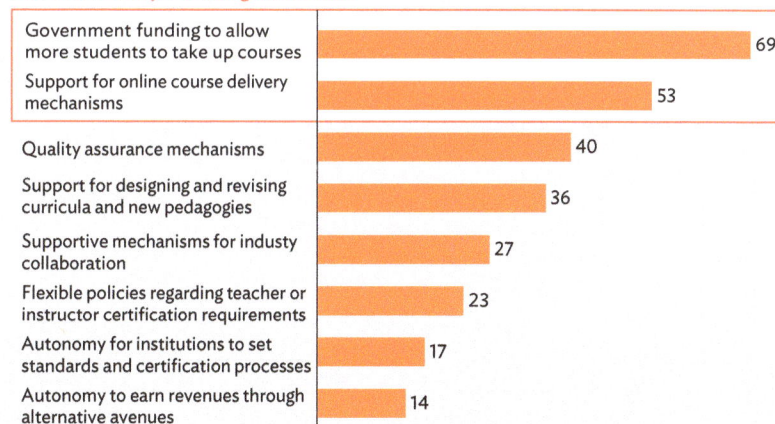

30% of training institutions find it at least somewhat difficult to fill vacancies

- 3% Extremely difficult
- 3% Difficult
- 24% Somewhat difficult
- 13% Somewhat easy
- 23% Easy
- 34% Extremely easy

Reasons for difficulties in filling places
Percent of surveyed training institutions answering "extremely difficult," "difficult," or "somewhat difficult"

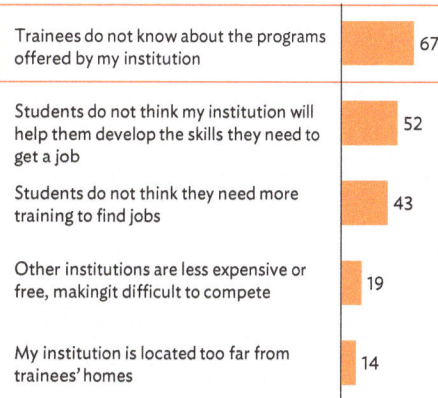

Trainees do not know about the programs offered by my institution	67
Students do not think my institution will help them develop the skills they need to get a job	52
Students do not think they need more training to find jobs	43
Other institutions are less expensive or free, makingit difficult to compete	19
My institution is located too far from trainees' homes	14

Note: Based on survey of training institutions between June and September 2021 (n=70). Percentages do not add up to 100% as respondents were asked to select all options that apply.
Source: Asian Development Bank (Sustainable Development and Climate Change Department).

Figure 42: Training Institutions' Perceptions on Most Impactful Public Policies for Training Provision in Azerbaijan (%)

Training institutions would like to see more support for students' course fees and for online course delivery mechanisms

Percent of surveyed training institutions

Government funding to allow more students to take up courses	69
Support for online course delivery mechanisms	53
Quality assurance mechanisms	40
Support for designing and revising curricula and new pedagogies	36
Supportive mechanisms for industy collaboration	27
Flexible policies regarding teacher or instructor certification requirements	23
Autonomy for institutions to set standards and certification processes	17
Autonomy to earn revenues through alternative avenues	14

Note: Based on survey of training institutions between June and September 2021 (n=70). Percentages do not add up to 100% as respondents were asked to select all options that apply.
Source: Asian Development Bank (Sustainable Development and Climate Change Department).

F.   Supply and Demand Mismatches

Insufficient information on job openings in the market was cited by training institutions as the most common reason for their graduates being unable to find jobs (Figure 43). Policies targeted at building a stronger awareness among students and training institutions of the jobs available with the adoption of 4IR technologies in key industries and the corresponding skills required could help to address the information asymmetry in the market.

There is also scope for training institutions to provide stronger support to graduates in their job search. Less than a third of training institutions surveyed organize industry visits and exchanges and only 37% provide job application support to students (Figure 44). This suggests that training programs are not focused on preparing workers for future employment in a large proportion of training institutions, which reinforces earlier findings that industry engagement is limited across training institutions in Azerbaijan. The lack of focus on career and professional development in training institutions was also highlighted as an issue by local experts and stakeholders at the country consultation workshops conducted. Students lack sufficient access to information on potential career paths and corresponding skill requirements. In contrast, 60% of training institutions surveyed provided scholarships to students from disadvantaged backgrounds, showing a strong focus on ensuring that socially disadvantaged communities have equitable access to training opportunities (Figure 44).

Figure 43: Training Institutions' Perception of Reasons for Students Being Unable to Find Jobs upon Graduation in Azerbaijan

The lack of information on job openings is a key barrier to graduates' job search

Ranking based on responses from surveyed training institutions;

1 - Most common; 5 – Least common

Rank

1	Students have insufficient information on job openings in the market
2	There are not enough job opportunities
3	Current job opportunities are not attractive enough to incentivize workers to complete relevant training programs
4	Education and training programs do not adequately prepare job seekers for jobs
5	The certifications provided to graduates of training institutions are not well-recognized by employers

Note: Based on survey of training institutions between June and September 2021 (n=70).
Source: Asian Development Bank (Sustainable Development and Climate Change Department).

Figure 44: Nontraining Initiatives Provided by Training Institutions to Support Trainees in Their Professional and Personal Development in Azerbaijan (%)

Only 37% of training institutions in Azerbaijan provide job application and interview support for their students

Percent of surveyed training institutions

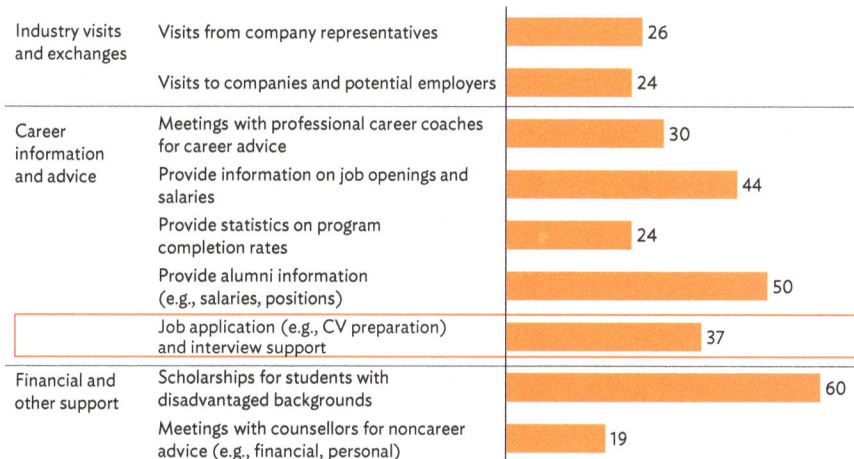

Category	Initiative	%
Industry visits and exchanges	Visits from company representatives	26
	Visits to companies and potential employers	24
Career information and advice	Meetings with professional career coaches for career advice	30
	Provide information on job openings and salaries	44
	Provide statistics on program completion rates	24
	Provide alumni information (e.g., salaries, positions)	50
	Job application (e.g., CV preparation) and interview support	37
Financial and other support	Scholarships for students with disadvantaged backgrounds	60
	Meetings with counsellors for noncareer advice (e.g., financial, personal)	19

Note: Survey of training institutions (n=70). Percentages do not add up to 100% as respondents were asked to select all options that apply.
Source: Asian Development Bank (Sustainable Development and Climate Change Department).

There are significant differences between the perceptions of training institutions and employers toward graduate quality in Azerbaijan. While 83% of training institutions surveyed feel that their graduates are adequately prepared for entry-level positions, only 18% of employers in the agro-processing industry and 24% of employers in the transportation and storage industry agree (Figure 45). Similar differences were observed when training institutions and employers were asked if graduates had the appropriate general and job-specific skills. There could be a few reasons for these differences in perceptions. First, training institutions could have an outdated understanding of industry's skills needs so that while they produce graduates skilled in some areas, these skills might be deemed irrelevant by employers who require different skills sets and knowledge. Second, as found from the training institution surveys, training institutions lack instruments to regularly appraise the performance of instructors so that knowledge delivery in the classroom could be compromised by poor-quality instructors. Various other gaps in the curricula development and knowledge delivery process of training institutions could explain the difference in perceptions. It is critical to address these gaps by ensuring stronger alignment between industry and training institutions on skills needs, and building strong mechanisms for quality assurance in the education system.

Figure 45: Perception of Employers on Graduates' Preparedness for Entry-Level Positions in Azerbaijan (%)

Training institutions perceive graduates to be better prepared for the workforce than employers

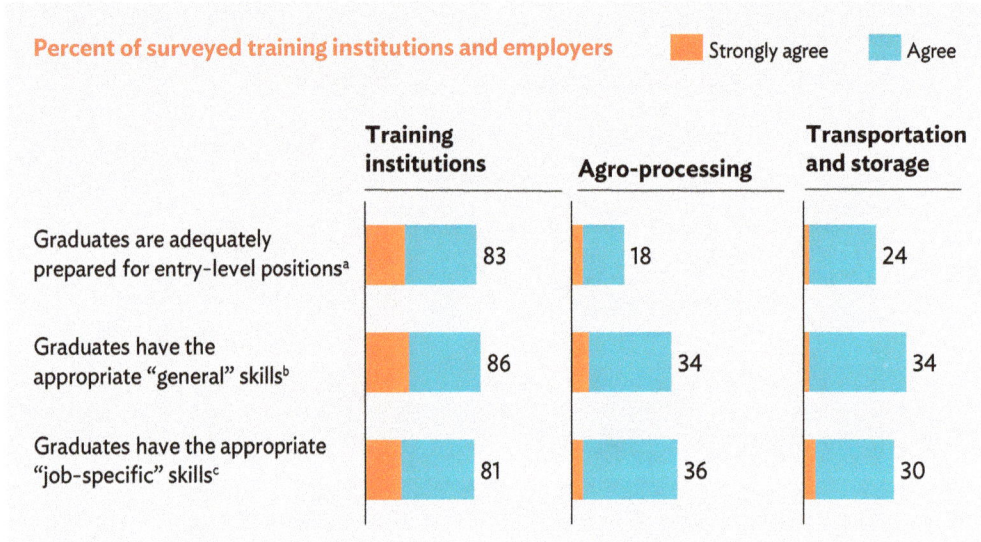

Percent of surveyed training institutions and employers   ■ Strongly agree   ■ Agree

	Training institutions	Agro-processing	Transportation and storage
Graduates are adequately prepared for entry-level positions[a]	83	18	24
Graduates have the appropriate "general" skills[b]	86	34	34
Graduates have the appropriate "job-specific" skills[c]	81	36	30

[a] In their chosen fields of study.

[b] "General" skills include soft and generic skills that are developed through any academic program and experience and are requisite for success in any job, e.g., teamwork, creativity, and problem solving.

[c] "Job-specific" skills include skills relevant to the discipline or job description that are necessary to succeed in that specific position, e.g., accounting, computer programming, and engineering.

Note: Based on survey of training institutions between June and September 2021 (n=70) and employer surveys (n=50 for agro-processing; n=50 for transportation and storage) between June and September 2021.

Source: Asian Development Bank (Sustainable Development and Climate Change Department).

3 National Policy Responses

The Government of Azerbaijan recognizes the potential of 4IR technologies in promoting growth, as demonstrated by the establishment of the Center for Analysis and Coordination of the Fourth Industrial Revolution in early 2021. Overall, the current degree of focus varies across the 4IR-relevant policy areas in Azerbaijan. There is a strong focus on the creation of new, flexible qualification pathways. The National Qualifications Framework for Lifelong Learning enables the recognition of skills gained through nonformal and informal education and creates a single standard against which education and vocational training, as well as work-based learning, can be measured. However, there is scope to consider incentives for workers and employers to participate in skills development, as well as strengthen social protection mechanisms for workers taking on flexible forms of labor. In terms of implementation, existing national plans acknowledge the emergence of new, advanced technologies, but do not set out a vision for 4IR adoption or road map to integrate 4IR trends with skills development across the economy. There also does not appear to be a coordinated approach toward 4IR adoption in Azerbaijan aligned across all stakeholders and backed by strong public financing. As such, the Centre for Analysis and Coordination of the Fourth Industrial Revolution marks a positive step toward addressing the current gaps in policy coverage and implementation and creating a coordinated policy approach that would allow Azerbaijan to be better prepared to leverage 4IR technologies for inclusive socioeconomic development.

The policy assessment leverages a combination of government policy documents, academic literature on Azerbaijan's skills development landscape and relevant government policies, as well as the surveys and skills gap analysis conducted as part of this country report.

A.   Overview of Industry 4.0 Policy Landscape

The Center for Analysis and Coordination of the Fourth Industrial Revolution under the Ministry of Economy, a new public entity tasked to coordinate Azerbaijan's response to 4IR trends, was established in early 2021 (Azerbaijan Press Agency 2021). National strategy documents reflect an intention to strengthen the country's human capital in relation to technology adoption. In 2016, Azerbaijan launched a series of strategic road maps that set out the government's plans for inclusive economic development for 2025 and beyond (Government of the Republic of Azerbaijan 2016f). These include the Strategic Roadmap on the National Economy Perspective, which sets out Azerbaijan's overall economic growth priorities, and 11 sector road maps. The sector road maps focus on accelerating growth in priority industries, including the agriculture and transportation and storage industries, through technology adoption and skills development. Azerbaijan's key 4IR and skills-related government policies include the following:

(i)   **Establishment of the Center for Analysis and Coordination of the Fourth Industrial Revolution.**
Established in early 2021, the center is tasked with strengthening the position of Azerbaijan in responding to 4IR trends and aims to promote innovative technology and expedite digitalization. It also

functions as the Azerbaijan Affiliate (C4IR Azerbaijan Affiliate) of the Center for the Fourth Industrial Revolution Network of the World Economic Forum. The principal goal of the C4IR Network of the World Economic Forum is to rapidly adopt new advanced technologies and develop the application and management of these technologies to serve citizens and the public interest.[14]

(ii) **Azerbaijan 2030: National Priorities of Socio-Economic Development and Socio-Economic Development Strategy for 2021–2025.** The Azerbaijan 2030: National Priorities for Socio-economic Development was approved in February 2021 and the Socio-Economic Development Strategy for 2021–2025 being developed will set out plans for Azerbaijan's inclusive growth and digital transformation. Two priorities in the strategy are particularly relevant for 4IR: (a) a steadily growing, competitive economy, and (b) competitive human capital and modern innovation space (Azertag 2021).

(iii) **Strategic Roadmap for the National Economy Perspective of the Republic of Azerbaijan.** This road map provides a vision for Azerbaijan's transition toward a knowledge-based economy by 2025, supported by a quality education and vocational training system and the adoption of advanced technologies. It outlines plans to improve the quality of education and develop a skilled, employable workforce (Government of the Republic of Azerbaijan 2016d).

(iv) **Strategic Roadmap for the Development of Telecommunications and Information Technologies in the Republic of Azerbaijan.** The road map sets out national plans to strengthen Azerbaijan's ICT infrastructure and leverage innovative technologies in all business sectors to increase competitiveness. It also highlights the need for ICT skills training for current employees and the future workforce (Government of the Republic of Azerbaijan 2016e).

(v) **Strategic Roadmap for Vocational Education and Training in the Republic of Azerbaijan.** The road map sets out actions that Azerbaijan intends to undertake to improve its VET system, including encouraging stronger employer involvement in VET, establishing a strong performance-based funding system for VET, and industry-specific training programs, as well as implementing mechanisms to recognize the competencies acquired through nonformal and informal education among other areas (Government of the Republic of Azerbaijan 2016c).

(vi) **National Qualifications Framework for Lifelong Learning in the Republic of Azerbaijan.** Adopted in 2018, this framework provides a basis for the recognition of skills obtained through formal, nonformal, and informal modes of education, and serves to facilitate comparability between national and international qualifications (ETF 2021b).

(vii) **National Employment Strategy 2019–2030.** The strategy sets out plans for increasing the productivity of the workforce through skills development and expanding employment and entrepreneurship opportunities. It also sets out plans to achieve inclusive growth by increasing the coverage and efficiency of labor market programs (ADB 2020).

(viii) **Strategic Vision and Roadmap for Azerbaijan Agriculture.** The vision and road map set out plans to increase the competitiveness of Azerbaijan's agricultural production and processing industries by encouraging private sector investment into agro-processing value chains and leveraging advanced technologies (Government of the Republic of Azerbaijan 2016a).

(ix) **Strategic Roadmap for the Development of Logistics and Trade in the Republic of Azerbaijan.** The road map sets out Azerbaijan's long-term vision to transform into a regional logistics and trade hub by leveraging the use of data analytics technology to build intelligent systems to optimize warehousing and transportation operations (Government of the Republic of Azerbaijan 2016b).

Key policies relevant to managing the impact of 4IR on skills in Azerbaijan are also summarized in Table 6.

---

[14] Currently, the C4IR Network has 13 centers. Of these, nine are affiliated centers, while three are centers run by the forum and headquartered in San Francisco, United States.

Table 6: Key Policies Relevant to Managing the Impact of Industry 4.0 on Skills in Azerbaijan

Policy	Responsible Entity(ies)	Relevance
Establishment of the Center for Analysis and Coordination of the Fourth Industrial Revolution	Ministry of Economy	New public entity under the Ministry of Economy tasked to strengthen Azerbaijan's responsiveness to 4IR trends
Azerbaijan 2030: National Priorities of Socio-Economic Development and Socio-Economic Development Strategy for 2021–2025	Cabinet of Ministers	Two national priorities are particularly relevant for 4IR, (i) a steadily growing, competitive economy; and (ii) competitive human capital and modern innovation space
Strategic Roadmap for Development of Telecommunications and Information Technologies in the Republic of Azerbaijan	Working group led by the Ministry of Transport, Communication, and High Technologies, which also includes representatives from government agencies covering the areas of education, economy, finance, foreign affairs, health, justice, and innovation	National strategy for information and communication technology development. Sets out plans to improve technology literacy for businesses and workers
Strategic Roadmap for Vocational Education and Training in the Republic of Azerbaijan	Working group led by the Ministry of Education, including representatives from the State Agency for Vocational Education, the Ministry of Finance, the Ministry of Economy and the Ministry of Labor and Social Protection of Population	National strategy for vocational training. Sets out plans for building a skilled workforce able to meet the needs of the labor market
National Qualifications Framework for Lifelong Learning of the Republic of Azerbaijan	Ministry of Education	National qualifications framework that covers both initial and lifelong learning at all levels of education
National Employment Strategy 2019–2030	Ministry of Labor and Social Protection of the Population	Strategy document that sets out plans for pro-employment macroeconomic policy as well as initiatives to support decent work, improve labor productivity, and utilize labor resources fully

4IR = Fourth Industrial Revolution.

Sources: *Azerbaijan Press Agency*. 2021. Center for Analysis and Coordination of the Fourth Industrial Revolution under the Ministry of Economy Established. 6 January. https://apa.az/en/infrastructure/Center-for-Analysis-and-Coordination-of-the-Fourth-Industrial-Revolution-under-the-Ministry-of-Economy-established-339541; Government of the Republic of Azerbaijan. 2016. *Strategic Roadmap for Development of Telecommunications and Information Technologies in Azerbaijan Republic*. Baku. https://monitoring.az/assets/upload/files/6683729684f8895c1668803607932190.pdf; Government of the Republic of Azerbaijan. 2016. *Strategic Roadmap for Vocational Education and Training in Azerbaijan Republic*. Baku. https://monitoring.az/assets/upload/files/6381dda5389fb17755bbb732a9c7d708.pdf; European Training Foundation. 2021. *National Qualifications Framework – Azerbaijan*. Turin. https://www.etf.europa.eu/sites/default/files/document/Azerbaijan.pdf; European Training Foundation. 2018. *Azerbaijan Country Strategy Paper 2017-20: 2018 Updates*. Turin. https://www.etf.europa.eu/sites/default/files/m/96B9F9A8EA4FF770C125821F005391A7_CSP%202017-2020%20AZERBAIJAN_Updates%202018.pdf.

B.   Assessment of Current Policy Approaches in Azerbaijan Related to Industry 4.0 and Skills

A diagnostic approach was taken to understand two important aspects of Azerbaijan's 4IR skills policy approach: (i) "the what"— specific policies being adopted by Azerbaijan and how they compare to international best practice approaches in preparing workers for 4IR; and (ii) "the how"—implementation mechanisms supporting 4IR efforts in government.

Assessment of Policy Actions ("The What")

Azerbaijan's policies and programs have been grouped into three action agendas that are assessed to be most crucial to managing the impact of 4IR on jobs and skills.[15] Figure 46 shows the current degree of focus by the country for each action area, rated as "strong," "moderate," or "weak" based on the analyzed extent of the policies' coverage in terms of scope and scale, as compared to those observed in international best practices. :

Figure 46: Degree of Focus of Policy Actions to Manage the Impact of Industry 4.0 on Jobs and Skills in Azerbaijan

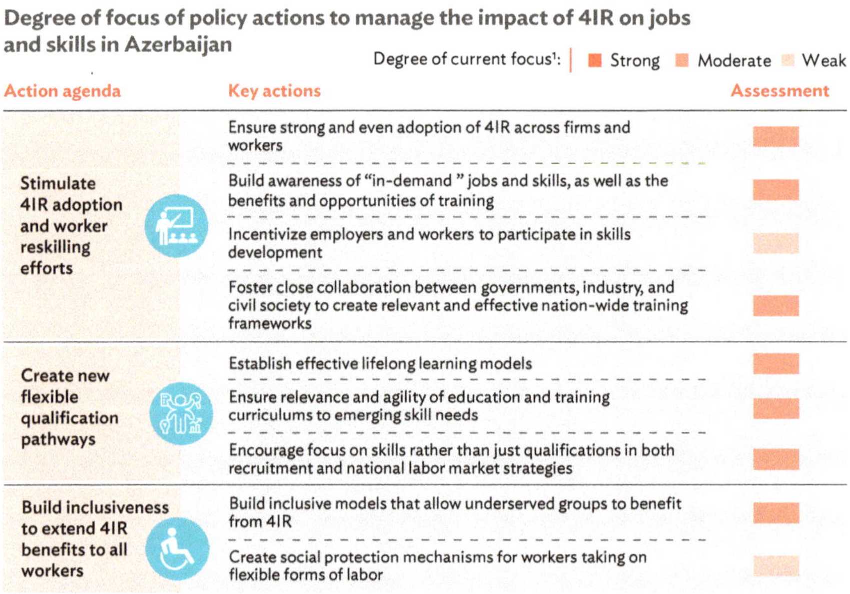

4IR = Fourth Industrial Revolution.

Note: Degree of focus was assessed based on the following criteria:

"Strong:" few or no gaps between the country's coverage of policy actions and coverage seen in international best practices;

"Moderate:" medium level of gaps between the country's coverage of policy actions and coverage seen in international best practices; and "Weak:" significant gaps between the country's coverage of policy actions and coverage seen in international best practices.

Sources: Asian Development Bank (Sustainable Development and Climate Change Department) and AlphaBeta.

---

[15]   Based on AlphaBeta research on international best practices for policy actions that manage the impact of Industry 4.0 on jobs and skills. For details of these best practices, see Microsoft and AlphaBeta (2019).

Overall, the current degree focus varies across the 4IR-relevant policy areas in Azerbaijan. There is a strong focus on the creation of new, flexible qualification pathways through building effective lifelong models and encouraging a focus on skills, rather than qualifications, in the labor market. However, there is scope to expand policy coverage to include incentives for workers and employers to participate in skills development, as well as strengthen social protection mechanisms for workers taking on flexible forms of labor. More specific steps are listed below.

(i) **Stimulating 4IR adoption and worker reskilling efforts.** Azerbaijan has adopted various strategies to encourage firms to adopt digital technology and upskill workers digitally. For instance, the Innovation Agency of Azerbaijan provides trained and mentored entrepreneurs in building their own technology start-ups and supported small and medium-sized enterprises (SMEs) in creating platforms for online sales during the COVID-19 pandemic (AzerNews 2020). The Small and Medium Business Development Agency trains SMEs to improve digital competencies and move toward a digital economy through its SME development centers established around the country.[16] The Ministry of Transport, Communications and High Technologies also launched an online portal to provide free training on basic digital skills and other ICT-related topics to all Azerbaijanis (AzerNews 2021a). Efforts to foster innovation were also highlighted by local government and industry stakeholders during the country consultations conducted. Universities and private firms have created digital laboratories, such as Microsoft's Internet of Things Laboratory that aims to create a bridge between innovation and implementation of new technological solutions.[17] The State Agency for Public Services and Social Innovations launched incubation programs that helped students to develop start-ups.[18] Nonetheless, there are various areas in which Azerbaijan could strengthen policy coverage to further stimulate 4IR adoption and working reskilling efforts:

   (a) **Strengthen focus on policies to educate businesses and workers on the benefits of 4IR technologies.** Around 14% of firms surveyed in the agro-processing industry and 20% of firms surveyed in the transportation and storage industry have limited knowledge of 4IR technologies despite firms in both industries having strong expectations of the potential labor productivity gains from adopting such technologies. For workers, current training programs to build basic digital literacy could be leveraged to focus on 4IR and knowhow for emerging technologies such as AI and IOT.

   (b) **Strengthen policies to build awareness of "in-demand" jobs and skills.** The European Training Foundation (ETF 2020b) and Government of the Republic of Azerbaijan (2016c) note that the establishment of the National Observatory on Labor Market and Social Protection Affairs to collect and analyze information on jobs and skills supply and demand, as well as sectoral committees that provide a platform for collaboration between government agencies, training institutions, and industry, are positive steps toward building stronger alignment in skills development needs among various stakeholders. However, past research indicates a significant level of skills mismatch in Azerbaijan, with workers being underqualified or overqualified for their roles, or not working within the specialization in which they were trained (ADB 2020, Government of the Republic of Azerbaijan 2019). As such, targeted policies to bridge the skills mismatch in Azerbaijan need to be developed.

   (c) **Focus on policies to incentivize firms to send workers for 4IR-related training and initiatives to build awareness of such policies or programs.** Only 32% of employers in the agro-processing industry and 34% of employers in the transportation and storage industry agree that they currently invest sufficiently in worker training.

---

[16] Country consultation workshop held in May 2021.

[17] Sources: Ministry of Transport, Communication and High Technologies. First Internet of Things Laboratory Created in Azerbaijan. https://mincom.gov.az/en/view/news/590/first-internet-of-things-laboratory-created-in-azerbaijan; and country consultation workshop.

[18] Country consultation conducted in May 2021.

(ii)   **Creating new flexible qualification pathways.** Azerbaijan has focused on the creation of new, flexible qualification pathways. The National Qualifications Framework for Lifelong Learning enables the recognition of skills gained through nonformal and informal education and creates a single standard against which education and vocational training, as well as work-based learning, can be measured (ETF 2021b). There are some efforts to make training curricula more relevant to new skills needs but policy focus in this area needs to be strengthened. Since 2018, Azerbaijan has started to develop and shift away from the old Soviet curricula toward modern curricula for vocational education designed to meet labor market needs with programs formulated based on occupational standards (ETF 2020a). A digital skills pilot program in secondary schools aims to introduce coding and programming languages as a part of the curriculum. Incubation programs aimed at mentoring students to develop their own start-ups were also launched (ETF 2020a). However, the employer surveys revealed that firms in the agro-processing and transportation and storage sectors continue to face challenges in hiring workers with the required general or "job-specific" skills sets. Similar concerns were raised by industry and government stakeholders during the country consultations, and stronger industry involvement in the design of training programs was highlighted as a possible means to help address the gap. Around half of firms surveyed in Azerbaijan's agro-processing industry and transportation and storage industries have no active partnerships with training providers but are open to pursuing such partnerships.

(ii)   **Building inclusiveness to extend 4IR benefits to all workers.** A higher proportion of net job gains from the adoption of 4IR technologies in both the agro-processing and transportation and storage industries is expected to go to male workers. This is because most new jobs are expected to be in the technical stream, which have a larger concentration of male workers in both industries. Thus, targeted policy action is necessary to ensure that women also benefit from 4IR. Some existing programs in Azerbaijan encourage women to pursue technical careers. One example is the Digital Girls national education program that trains girls in digital skills and computer technology and educates them on building a successful ICT career. The UNDP also has a mentorship program for young female professionals and students in Azerbaijan keen to pursue a career in STEM (UNDP Azerbaijan 2021). However, these programs could be further broadened and could focus more strongly on the specific impact of adopting 4IR technologies and related skills needs. In addition, social protection and skilling policies would need to be updated to address new types of workers that could emerge in a knowledge-based economy, including digital freelancers or gig economy workers.

Assessment of Implementation of Industry 4.0 Policies ("The How")

The implementation of Azerbaijan's 4IR strategy for jobs and skills was assessed against three dimensions found to be crucial for success according to past academic work: (i) clarity and robustness of plans, (ii) strength of coordination between different stakeholders, and (iii) alignment of financing and incentives (Figure 47).[19]

There is no coordinated approach toward 4IR adoption in Azerbaijan that is aligned across all stakeholders and backed by strong public financing. Nevertheless, overall, there is scope to strengthen the implementation of 4IR policies, more specifically through the following measures:

(i)   **Clarity and robustness of plans.** The Strategic Roadmap on the National Economy Perspective provides a vision for Azerbaijan's transition toward a knowledge-based economy by 2025. It has 11 sectoral road maps, including road maps covering the agro-processing and transportation and storage industries, and sets out plans for technology adoption and skills development in key growth sectors. However, there are two ways in which the clarity and robustness of Azerbaijan's 4IR plans could be

---

[19]   Based on AlphaBeta research of Industry 4.0 strategies, plus insights from past public sector research, including Barber (2007) and McKinsey & Company (2012).

Figure 47: Implementation Challenges Associated with Industry 4.0 Policies
for Jobs and Skills in Azerbaijan

Implementation challenges associated with 4IR policies for jobs and skills in Azerbaijan

Degree of current focus[1]: ■ Strong ■ Moderate ■ Weak

Dimension	Questions	Assessment
Clarity and robustness of plans	Is there a clearly articulated vision for 4IR?	Weak
	Is there strong integration between employment or skills and the 4IR plan?	Weak
	Is the plan forward-looking, incorporating 4IR trends?	Weak
	Is there strong local data to support evidence-based policymaking?	Moderate
Strength of coordination	Is there one shared road map across industry and government departments for 4IR?	Weak
	Is there coordination across different government ministries and levels?	Moderate
	Is there strong alignment within and between industry, and education and training institutions?	Moderate
Alignment of financing and incentives	Is government financing aligned with the strategic goals?	Moderate
	What is the strength of incentives for employers and workers to invest in skill development? What is the strength of incentives for teachers and institutions to ensure high-quality training and education systems?	Moderate

4IR = Fourth Industrial Revolution.
Note: Degree of focus was assessed based on the following criteria:
"Strong:" few or no gaps between the country's policy implementation approach and approach seen in international best practices.
"Moderate:" medium level of gaps between the country's policy implementation approach and approach seen in international best practices.
"Weak:" significant gaps between the country's policy implementation approach and approach seen in international best practices.
Sources: Asian Development Bank (Sustainable Development and Climate Change Department) and AlphaBeta.

strengthened. First, explicitly address how 4IR technologies can be adopted and consider the skills development needs related to adoption. The establishment of the Center for Analysis and Coordination of the Fourth Industrial Revolution marks a positive step in including 4IR in the local policy discourse, and the drafting of a clear 4IR plan could be one of the Center's early priorities (Azerbaijan Press Agency 2021). Second, access to strong local data to support evidence-based policy making must be a priority. The creation of the National Observatory on Labor Market and Social Protection Affairs marks a positive step in collecting information related to jobs and skills supply and demand, and efforts could be taken to ensure that data collected will also be relevant to the design of 4IR-related jobs and skills strategies (ETF 2020b).

(ii) **Strength of coordination.** There are some efforts to improve coordination among government agencies and between training institutions and industry stakeholders on industry development and skills development plans. However, there are various ways in which coordination could be strengthened. As existing industry road maps do not specifically consider how to enable the adoption of 4IR technologies and the skills development needs arising from adoption, the role that each stakeholder must play in enabling the transition toward 4IR must first be clarified. Second, ensure that the newly established

Center for Analysis and Coordination of the Fourth Industrial Revolution coordinate the various aspects of innovation, technology adoption, and skills development in Azerbaijan.[20] Third, strengthen alignment between industry and training institutions. Sectoral committees that provide a platform for collaboration between government agencies, training institutions, and employers to develop occupational and qualifications standards have been established (ETF 2014). To date however, there is limited engagement between employers and training institutions. Only 20% of training institutions surveyed gather input for curricula from industry stakeholders and only 26% offer teaching placements for industry professionals at training institutions.

(iii) **Alignment of financing and incentives.** The share of education in total government spending in Azerbaijan fell from 24% in 2000 to less than 11% in 2020, which is below the 12% average of countries in Europe and Central Asia (ADB 2020, Center for Economic and Social Development 2020). Efforts to align the incentives of stakeholders to human capital development goals could still be improved. Teachers' salaries in VET institutions have been increased more than twice since 2018, and there are plans to implement a performance-based incentive system for vocational institutions and teachers that could help to improve the quality of training (ETF 2020a). During the country consultations, local stakeholders indicated that corporate tax exemptions of up to 10% of profit are provided to companies to invest in institutions that operate in the fields of science and education, although few companies are aware of the provision.[21] However, the absence of incentives for businesses to send workers for training means that many firms do not actively upskill their workers. Incentives should be strengthened to ensure that all stakeholders are aligned in the establishment of a quality skills development system to support 4IR reskilling needs.

Assessment of Industry 4.0 Policies in Relation to the COVID-19 Pandemic

Expectations on the impact of COVID-19 on technology adoption are muted in Azerbaijan, with only around a quarter of firms in both industries agreeing that COVID-19 has accelerated or will accelerate the adoption of 4IR technologies. As indicated in Chapter 1, two reasons could explain this. First, firms have a limited understanding of 4IR technologies and their applications. Second, COVID-19 has created uncertainty among firms on their future economic prospects, which therefore leads to unwillingness to invest in adopting 4IR technologies.

Notwithstanding the uncertainty among firms on the impact of COVID-19 on technology adoption, the government has put in place various initiatives to facilitate the digital transformation of businesses and upskill the workforce. The Azerbaijan Small and Medium Business Development Agency, through its Small and Medium Business Development Centers, regularly provides training and consulting services on various topics (including digitalization) and implements outreach activities on support schemes available to support small businesses in Azerbaijan.[22] The agency has embarked on various efforts to improve awareness of digital business practices, including through training sessions and webinars on the use of e-commerce platforms and digital marketing. The agency also developed a video training platform (www.kobim.az) to provide training in various areas and was tasked to develop an online business portal for SMEs (footnote 22). Support has also been provided to SMEs to create platforms for online sales during the pandemic, and an online sales platform and e-trade site (www.kobmarket.az) to facilitate the promotion of Azerbaijani products has been launched (AzerNews 2020). In addition, the Ministry of Labor and Social Protection, State Employment Agency, UNDP, and online training platform Coursera collaborated to launch the Workforce Recovery Initiative to provide training for 50,000 Azerbaijanis who lost their jobs due to COVID-19, to upskill their knowledge and competences. The initiative

---

[20] Country consultation workshops.
[21] Sources: BDO Tax News. Azerbaijan Recent Tax Changes. https://www.bdo.global/en-gb/microsites/tax-newsletters/corporate-tax-news/issue-50-february-2019/azerbaijan-recent-tax-changes; and country consultations conducted in September 2021.
[22] Inputs from government stakeholders; country consultations.

offers an online platform to help unemployed workers gain new skills and competencies to re-enter the labor market (UNDP Azerbaijan 2020a). The Ministry of Transport, Communications and High Technologies, in cooperation with UNDP, launched the one-stop digital platform "Stay Home," which, among other services, provided e-learning resources on how to set up and run digital businesses from home (UNDP Azerbaijan 2020b).

Beyond pursing efforts to support affected Azerbaijani firms and workers during the pandemic, the government would also need to consider policies to encourage firms and workers to adopt technology in the long term, so that they will remain competitive. Based on a global survey of executives, the COVID-19 pandemic has led to firms accelerating the digitization of their customer and supply-chain interactions and their internal operations by 3–4 years (McKinsey & Company 2020). Azerbaijani firms must similarly adopt digital technologies to remain globally competitive. In Singapore, the Productivity Solutions Grant supports firms keen on adopting ICT solutions and equipment to enhance business processes by subsidizing the cost of adopting a suite of approved solutions, including tools for customer management, data analytics, financial management, and inventory tracking.[23]

---

[23] Enterprise Singapore. Productivity Solutions Grant. https://www.enterprisesg.gov.sg/financial-assistance/grants/for-local-companies/productivity-solutions-grant.

4   The Way Forward

The previous three chapters highlight the potential benefits that 4IR can bring to Azerbaijan as well as the challenges that would need to be addressed to achieve these benefits. This chapter identifies policy recommendations, based on global best practices, which policy makers in Azerbaijan can consider adopting, to unleash the potential opportunities created by Industry 4.0.

A.   Recap of Industry 4.0-Related Challenges Facing Azerbaijan

Table 7 provides a recap of the challenges facing Azerbaijan from the industry analysis (Chapter 1), the training institution survey (Chapter 2), and the policy assessment (Chapter 3).

Table 7: Recap of Challenges Facing Azerbaijan in Relation to Industry 4.0

Area	No.	Key Challenges	Findings
Agro-processing industry	1	Limited understanding of 4IR technologies and their applications	Apart from IOT technologies, adoption of 4IR technologies is expected to remain low among agro-processing firms by 2025
	2	Lack of good quality training providers	Over **60%** of firms disagree that it is easy to find good quality trainers
	3	Job gains from 4IR benefit male workers more than female workers	The number of new jobs expected to be gained by male workers is **1.6 times** that expected to be gained by female workers
Transportation and storage industry	4	There is limited understanding of the applications of 4IR technologies among firms	Only **20%** of transportation and storage firms have a good understanding of 4IR technologies and applications
	5	Most jobs created will be in technical roles that the workforce might not be equipped for	**32%** of transportation and storage firms expect the number of technical jobs to increase
	6	Significant changes in skill demand may lead to challenges in hiring workers	Digital and/or information and communication technology skills, creative thinking and/or design, and numeracy skills will become more important with 4IR adoption

continued on next page

Table 7 *continued*

Area	No.	Key Challenges	Findings
Training institutions	7	Limited collaboration with employers on training curricula	Less than half of employers provide input to training institutions to incorporate the latest industry knowledge in their training curricula
	8	Few training institutions provide courses on 4IR in 2020	Over half of firms plan to adopt IOT technologies but only **4%** of training institutions offer relevant courses
	9	Training institutions lack resources to prepare workers for 4IR	About **54%** of training institutions strongly agree that technical and financial support is needed to enable them to prepare workers for 4IR
Policy assessment	10	Lack of clearly articulated 4IR vision integrated with jobs and skills	Plans to improve education and vocational training systems do not incorporate 4IR trends
	11	Lack of incentives for firms and workers to build digital skills	About **69%** of training institutions would like to see more government funding to allow students to take up courses
	12	Lack of inclusive skilling opportunities and social protection mechanisms for flexible workers	Limited focus on social protection policies to address new types of workers that could emerge in a knowledge-based economy, including digital freelancers or gig economy workers

4IR = Fourth Industrial Revolution, IOT = Internet of Things.
Sources: Asian Development Bank (Sustainable Development and Climate Change Department) and AlphaBeta.

B.   Recommendations to Address Challenges

There are several areas in which Azerbaijan can strengthen its approach to 4IR to address the challenges outlined above. This section provides policy recommendations for policy makers, drawing from international best practices, as summarized in Figure 48. Table 8 provides a summary of recommendations and potential lead agencies in Azerbaijan to implement each recommendation as well as the approximate implementation timeframe.

Table 8: Summary of Recommendations, Potential Lead Agencies, and Approximate Time Frame for Implementation

Recommendations	Potential Lead Agency	Approximate Timeframe for Implementation
1. Develop 4IR adoption plans to complement existing sectoral road maps	Center for Analysis and Coordination of the Fourth Industrial Revolution	12–36 months
2. Develop programs and 4IR competency centers to build awareness of digital tools among firms	Center for Analysis and Coordination of the Fourth Industrial Revolution	Less than 12 months
3. Implement incentive schemes for firms to train employees for 4IR	Small and Medium Business Development Agency of Azerbaijan	12–36 months
4. Programs to strengthen industry knowledge and digital skills of trainers and teachers	Ministry of Education	12–36 months

continued on next page

Table 8 *continued*

Recommendations	Potential Lead Agency	Approximate Timeframe for Implementation
5. Develop online learning platforms	Ministry of Education, State Agency for Public Services and Social Innovations	Less than 12 months
6. Develop innovative job-matching initiatives and platforms	Ministry of Labor and Social Protection of the Population, State Agency for Public Services and Social Innovations	Less than 12 months
7. Develop skilling and labor support programs for digital freelancers	Innovation Agency of Azerbaijan, Ministry of Transport, Communication and High Technologies	12–36 months
8. Develop programs to enable more women to take up technical jobs	Center for Analysis and Coordination of the Fourth Industrial Revolution	12–36 months

4IR = Fourth Industrial Revolution.

Sources: Asian Development Bank (Sustainable Development and Climate Change Department) and AlphaBeta.

Figure 48: Relevant Best Practices That Could be Adopted to Tackle Challenges in Adoption of Industry 4.0 Practices

There are a range of relevant best practices that could be adopted to tackle these challenges

- Address agro-processing industry challenges
- Address training landscape challenges
- Address transport industry challenges
- Address policy assessment challenges

Recommendations	Challenges addressed	Economies with best practices and/or similar solutions
1. Develop 4IR adoption plans to complement existing sectoral road maps	▪ Lack of clearly articulated 4IR vision integrated with jobs and skills aligned across stakeholders Lack of alignment between industry and training institutions on skills needs	Singapore
2. Develop programs and 4IR competency centres to build awareness of digital tools among firms	▪ Limited understanding of 4IR technologies among firms	Germany, Singapore,
3. Implement incentive schemes for firms to train employees for 4IR	▪ Lack of incentives for firms and workers to build digital skills	Japan, Malaysia, Singapore
4. Programs to strengthen industry knowledge and digital skills of trainers and teachers	▪ Training institutions lack resources to prepare workers for 4IR	Hong Kong, China; Malaysia; United Kingdom,
5. Develop online learning platforms	▪ Limited opportunities for adult learning	Republic of Korea
6. Develop innovative job-matching initiatives and platforms	▪ Significant changes in skill demand may lead to challenges in hiring workers	India, Malawi, Singapore
7. Develop skilling and labor support programs for digital freelancers	▪ Lack of inclusive skilling opportunities and social protection mechanisms for flexible workers	Pakistan
8. Develop programs to enable more women to take up technical jobs	▪ Job gains from 4IR benefit male workers more than female workers	Australia , Indonesia, Philippines, Thailand

4IR = Fourth Industrial Revolution.

Sources: Asian Development Bank (Sustainable Development and Climate Change Department) and AlphaBeta.

Recommendation 1: Develop Industry 4.0 Adoption Plans to Complement Existing Sectoral Road Maps

The Strategic Roadmap on the National Economy Perspective provides a vision for Azerbaijan's transition toward a knowledge-based economy by 2025. Eleven sector road maps, including road maps covering the agro-processing and transportation and storage industries, set out plans for technology adoption and skills development in key growth sectors. However, these plans do not explicitly consider the impact of 4IR trends on industry in terms of tasks and job impacts, nor do they consider the specific skills that will be needed.

Participants at the country consultation workshops stressed the need to align plans and policies for firms to adopt 4IR closely with skills development policies. In this area, the Industry Transformation Maps (ITMs) developed by the Government of Singapore could provide a useful model. Industry-specific ITMs have been developed for 23 industries, drawing together the inputs from private and public stakeholders in each industry, including trade associations and key firms. Each ITM charts out the overall growth direction for the industry under different pillars of transformation, such as jobs and skills, productivity, innovation, and internationalization. For example, the Logistics ITM sets out Singapore's vision of building a global leading logistics hub underpinned by operations excellence, innovation, and a strong talent pool. Under the productivity and innovation pillars, it sets out the government's overall goals to support firms to adopt new, advanced technologies and bolster the logistic innovation ecosystem in Singapore by working with research institutions. The jobs and skills or talent pillar sets the objectives of creating of quality jobs within the industry and enabling workers to transit smoothly into new, higher-skilled roles enabled by technology adoption. Specific policies and programs are guided by these broad objectives aligned across stakeholders.[24] Unlike the current strategic road maps in Azerbaijan, Singapore's ITMs explicitly consider how innovation and technology will transform jobs and skills in each industry. For instance, while Azerbaijan's road map for logistics mentions plans to implement advanced technological infrastructure, Singapore's logistics ITM sets out specific technologies and projects that will be implemented (e.g., integrated goods mover systems). In addition, while both sets of plans provide a projection of the number of jobs expected to be created in the logistics industry, Singapore's ITM specifically considers the change in the types of jobs available, and by extension the skilling requirements of the workforce, with the adoption of advanced technologies whereas the existing road map in Azerbaijan does not.

The newly established Center for Analysis and Coordination of the Fourth Industrial Revolution could take the lead in coordinating these plans in Azerbaijan, working with key line agencies. For instance, an action plan for 4IR technology adoption in the agro-processing industry could be cochaired by the Center and Ministry of Agriculture. Policies related to industry development and skills development often span a wide range of government stakeholders. A strong coordinating agency is needed to coordinate these efforts and resolve any differences between stakeholders, as well as to incorporate the needs of the private sector. A clear mandate must therefore be given to the center. For instance, Singapore's Logistics ITM is led by the Economic Development Board, the lead agency for the sector, but the overall implementation is overseen by the Future Economy council chaired by the deputy prime minister.[25]

Recommendation 2: Develop Programs and Industry 4.0 Competency Centers to Build Awareness of Digital Tools among Firms

There is limited understanding of 4IR technologies among firms in Azerbaijan. Without targeted efforts to increase awareness of 4IR technologies relevant to each industry among firms, and without specific digital tools, net job gains will be limited. In addition, over half of all enterprises in Azerbaijan are small enterprises with

24   Ministry of Trade and Industry Singapore. ITMs Logistics. https://www.mti.gov.sg/ITMs/Trade_Connectivity/Logistics.
25   Ministry of Trade and Industry Singapore. ITMs Overview. https://www.mti.gov.sg/ITMs/Overview.

fewer than 25 employees so capital investments in adopting 4IR are likely to be viewed as significant (Aliyev 2019). These firms would therefore require special support to understand the applications of 4IR technologies and benefits of adoption. The employer surveys revealed that only 14% of agro-processing firms and 20% of transportation and storage firms have a good understanding of 4IR technologies in 2020. A scan of the policy landscape in Azerbaijan also reveals that there is limited focus on programs that encourage firms, particularly those outside the ICT sector, to adopt 4IR technologies.

In Singapore, Industry Digital Plans have been created to guide firms, particularly SMEs, in the adoption of 4IR technologies. The plans provide clear guidance on the industry-specific digital tools that firms can adopt at different stages of growth. Diagnostic tools are available for firms to gauge their readiness to adopt various digital tools.[26] For instance, under the Logistics Industry Digital Plan, logistics firms at a nascent stage of technology adoption are guided to implement software to enable fleet or freight management, such as software that digitalize information flow in freight forwarding operations. Those at a more advanced stage are guided to implement augmented reality or virtual reality technologies for training and in areas such as warehouse redesign and planning.[27]

In Germany, the Mittelstand 4.0 Competence Centers was established with the support of the Federal Ministry of Economics and Technology to help businesses gauge their current stage of digitalization, and jointly develop and implement digitalization road maps and solutions (BMWI 2019). Operated by consortiums consisting of universities, research institutions, and industry chambers, these competence centers can provide information and training to SMEs on digital transformation; and operate demonstration plans where they show variants of new digital technologies in stimulated business and production processes so that SMEs can understand how these technologies would impact the operations in a real-world environment (BMWI 2019).

In Azerbaijan, the Center for Analysis and Coordination of the Fourth Industrial Revolution is developing programs to provide customized support for companies to adopt digital technologies, and a web portal has been established to provide local firms with information on how to digitalize their businesses. The firm support program would have a two-stage process, with the first stage focusing on providing a diagnostic of each firm's current stage of technology adoption and readiness for new technologies, and the second stage focusing on working with each firm to develop a customized technology adoption road map. These efforts could take reference from Singapore's Industry Digital Plans, which provide a tool kit for firms to do their own digital readiness assessment and a set of industry-specific technologies that they can adopt based on the results of the assessment. By guiding firms to conduct their own assessment, resources can be spread out to support a larger group of firms. As a start, Industry Digital Plans could be devised for the agro-processing and transportation and storage industries to identify the methodologies and frameworks that would be relevant to firms in Azerbaijan in diagnosing their readiness for digital transformation. These plans could then be expanded to cover more industries. The next step for the Center for Analysis and Coordination of the Fourth Industrial Revolution could be to develop 4IR competency centers modelled after the German model that can provide training to firms and demonstrate the impact that adopting specific technologies could have on their business. Azerbaijan could also expand these centers to provide 4IR-specific training courses to workers, co-developed with universities and other training institutions.

---

[26]   IMDA. *Industry Digital Plans.*https://www.imda.gov.sg/programme-listing/smes-go-digital/industry-digital-plans.

[27]   IMDA. *Logistics Industry Digital Plans.* https://www.imda.gov.sg/-/media/Imda/Files/Programme/SMEs-Go-Digital/Industry-Digital-Plans/Logistics-IDP/Logistics-IDP-08142020.pdf.

Recommendation 3: Implement Incentive Schemes for Firms to Train Employees for 4IR

Incentive schemes to encourage firms to provide training to workers could contribute to building a 4IR-ready workforce as well as reduce the risk of some groups of workers being displaced by automation. The employer surveys reveal that workers in manual or administrative roles within the agro-processing and transportation and storage industries are expected to face a higher risk of being displaced by 4IR while jobs are created in technical roles. These workers would require reskilling to take on new job roles and firms could help to provide the reskilling. However, only around a third of firms surveyed across both industries indicated that they currently invest sufficiently in training employees. Firms would therefore require stronger incentives to train employees in preparation for 4IR.

Policy makers in Azerbaijan could take reference from incentive schemes implemented in Japan, Malaysia, and Singapore (Box 3). In particular, incentives could be focused on SMEs that face the most significant resource constraints in sending workers for training, particularly as over half of all enterprises in Azerbaijan are small enterprises with fewer than 25 employees (Aliyev 2019). The Small and Medium Business Development Agency of Azerbaijan supports projects that enable SMEs to benefit from advanced technologies and apply relevant technologies in their business activities, including through the provision of experts.[28] The agency is well-placed to work closely with SMEs to identify the specific challenges they face in providing training to their workers and design programs to counter these challenges.

Box 3: Incentive Schemes for Employee Training in Japan, Malaysia, and Singapore

In Singapore, various forms of support are provided to employers to train their workers. Small and medium-sized enterprises (SMEs) that send their employees to attend training courses can benefit from the Enhanced Training Support for SMEs Program of Skills Future Singapore, which provides funding for up to 90% of course fees. To ensure that employers are not deterred by the loss of human resources when employees undergo training, the program also offers absentee payroll funding to cover up to 80% of the worker's salary during the training period. In addition, industry-specific job redesign reskilling and/or redeployment programs support firms that are undergoing business transformation, or which have existing workers at risk of redundancy or in vulnerable jobs due to the transformation. Support is provided to allow workers to be trained to take on new job roles or redesigned job roles within the company.

In Malaysia, a Skills Upgrading Program of SME Corp Malaysia provides grants covering 70% of training fees for technical and soft skills for SMEs to train their workers.

Meanwhile, the Organisation for Economic Co-operation and Development (OECD 2018) reports that in Japan, the *Jinzai Kaihatsu Shien Joseikin* (Subsidy to Support Human Resource Development) program subsidizes firms for their reimbursement of employees' wages during training, with the amount of subsidy being tailored to the type of training and size of the firm.

Source: Skills Future Singapore. Job Redesign Reskilling/Redeployment Programs. https://www.wsg.gov.sg/programmes-and-initiatives/job-redesign-reskilling-redeployment.html; SME Corp. SME Skills Upgrading Programme. https://www.fmm.org.my/images/articles/branches/Negeri_Sembilan/SME%20PROG.pdf; and OECD. 2018. *Getting Skills Right: Future-Ready Adult Learning Systems*. Paris. https://read.oecd-ilibrary.org/education/getting-skills-right-future-ready-adult-learning-systems_9789264311756-en#page1.

---

[28] Inputs from stakeholders; country consultations.

The Small and Medium Business Development Agency of Azerbaijan could also use Singapore's industry-specific Job Redesign Reskilling/Redeployment Program as a model to design programs that will support the reskilling of Azerbaijani workers in manual or customer-facing roles that face a higher risk of being displaced by 4IR, so that they can take on new jobs in technical roles or roles that require different skills. For instance, if companies face difficulties in engaging suitable trainers or lack the financial resources to pay for training, support could be provided to source suitable trainers or training programs, and provide course and payroll subsidies.

Recommendation 4: Develop Programs to Strengthen Industry Knowledge and Digital Skills of Trainers and Teachers

The quality of instructors is integral to any training and education system. To ensure that graduates gain relevant and practical skills needed to succeed in the workplace, it is critical that instructors have strong pedagogical knowledge as well as a good grasp of current industry trends. Less than half of training institutions surveyed in Azerbaijan currently provide professional development and training for instructors to update their knowledge of the latest equipment used and skills required by industry.

To ensure that instructors have updated skills and knowledge, industry stakeholders and training providers could collaborate on industrial attachments for instructors, or encourage experienced industry professionals to join their TVET workforce. In Malaysia, INTI International University & Colleges created the Faculty Industry Attachment program. The program enables teaching staff to work with industries as part of their regular working hours to broaden their practical experiences and stay abreast of the latest developments in the industry. Lecturers undergo an industrial attachment in related organizations for up to 96 hours (an estimated two-and-a-half working weeks), to gain an in-depth understanding of the current demands of the industry.[29] In the UK, the Taking Teaching Further program is a national initiative to attract experienced industry professionals with expert technical knowledge and skills to work in education. Training providers receive funding support to recruit people with industry experience to retrain for teaching positions. The funding covers the cost of undertaking a teaching qualification as well as the costs of inducting a new teacher into the system (i.e., work-shadowing arrangements and reduced workload for new teachers).[30]

Apart from industry knowledge, training institutions also need to improve the digital and ICT skills of their teaching staff and ensure that curricula and pedagogy are updated to meet the new skill needs created by 4IR. Firms in both the agro-processing and transportation and storage industries prioritized digital and/or ICT skills as the most important skill for employees in 2025 but training institutions surveyed indicated that their graduates have relatively low competency in this area. In Hong Kong, China, an ICT training framework has been developed (see Box 4) to ensure that teachers themselves possess sufficient digital skills so that they can pass these on to students (Fau and Moreau 2018). In the Republic of Korea, regular reviews of the national curriculum framework are undertaken by the country's Ministry of Education every 5–10 years (see Box 5), and the national Professional Development Master Plan, developed in 2015, lays out a comprehensive structure for professional learning throughout a teacher's teaching career (National Center on Education and the Economy 2020). The Ministry of Education of Azerbaijan could take the lead in programs to improve the competencies of teaching staff in schools and training institutions at various levels. Industry attachment programs and programs targeted at incentivizing industry professionals to join the teaching profession could be focused on vocational training institutions while programs to build strong digital literacy among teaching staff and ensure continued professional development could be implemented at all levels, to ensure that students in Azerbaijan form strong fundamentals in digital literacy from an early age.

---

[29] INTI International University & Colleges. *Industry Attachment Programme Redefines Teaching at INTI.* https://newinti.edu.my/industry-attachment-programme-redefines-teaching-at-inti/.
[30] Education and Training Foundation. Taking Teaching Further. https://www.et-foundation.co.uk/supporting/support-for-teacher-recruitment/taking-teaching-further/.

Box 4: Information and Communication Technology Training Framework for Teachers in Hong Kong, China

Hong Kong, China's information and communication technology (ICT) training framework was developed to ensure that instructors possess sufficient digital skills and can pass on those skills to their students (Fau and Moreau 2018). This framework seeks to achieve two outcomes: increase the digital competence of teachers and improve their teaching pedagogies for ICT skills in classrooms. The training framework has four dimensions that correspond to the range of skills that teachers will need to adapt their methods to a changing technological environment:

- technical skills in ICT,
- ICT teaching pedagogies,
- management and supervision of digital technology in classroom settings, and
- sociocultural awareness in online environments.

Considering the constant evolution of digital technologies, the training framework covers an initial training period for teachers, as well as continuing training. The trainings delivered to each teacher are also tailored to the specific requirements of preschool, primary, and secondary education, depending on the level taught by each teacher.

Source: S. Fau and Y. Moreau. 2018. *Managing Tomorrow's Digital Skills: What Conclusions Can We Draw from International Comparative Indicators?* United Nations Educational, Scientific and Cultural Organization (UNESCO) Digital Library. https://unesdoc.unesco.org/ark:/48223/pf0000261853.

Box 5: Ensuring that Curricula and Pedagogy are Updated to Meet New Skill Needs in the Republic of Korea

The Republic of Korea has one of the most tech-savvy student populations globally. This is reflected by its second-place ranking in the Organisation for Economic Co-operation and Development's Program for International Student Assessment (PISA) assessment in 2018 for "digital competence," which is defined by the ability to read, navigate, and understand online texts (PISA 2012).

A key driver behind this is the regular review every 5–10 years of the national curriculum framework by the country's Ministry of Education (National Center on Education and the Economy 2020). This policy seeks to ensure that educational curricula reflect updated learning needs in tandem with the emerging demands of the labor market including new technologies. A revision made in 2015 added six general competencies to students' learning outcomes, one of which is "21st century skills." This includes the introduction of both "soft" skills (such as creative thinking) as well as technical information and communication technology (ICT) skills that the education ministry deems would become increasingly important in the workforce (Kim et al. 2017). The 2015 revision, which had the target of being fully implemented by 2021, also led to the introduction of "creative experiential learning" activities. These are hands-on activities that encourage creative thinking, require mandatory coding classes for all elementary and middle school students, and include ICT training for teachers (Kim et al. 2017).

Apart from ensuring that curriculum is updated, strong emphasis is also placed on ensuring that teachers are updated on pedagogy. The national Professional Development Master Plan, developed in 2015, lays out a comprehensive structure for professional learning throughout a teacher's teaching career in the Republic of Korea. It recommends specific professional learning for educators according to their stage of career development and supports participation in training programs.

Sources: PISA. 2012. Main Results from the PISA 2012 Computer-Based Assessments. https://www.oecd-ilibrary.org/docserver/9789264239555-6-en.pdf?expires=1649834501&id=id&accname=guest&checksum=45BD7094B35E5CA72103B75AF6B7D1E3; National Center on Education and the Economy. 2020. South Korea: Learning Systems. http://ncee.org/what-we-do/center-on-international-education-benchmarking/top-performing-countries/south-korea-overview/south-korea-instructional-systems/; Kim et al. 2017. Korea's Software Education Initiative. https://dl.acm.org/doi/abs/10.1145/3151759.3151800?download=true.

Recommendation 5: Develop Online Learning Platforms

The adoption of 4IR technologies will rapidly and constantly change the skills and practical knowledge required of workers. Modern skills development systems would need to put in place platforms that would enable the constant renewal of skills in the workforce. In Azerbaijan, the National Qualifications Framework for Lifelong Learning enables the recognition of skills gained through nonformal and informal education and creates a single standard against which education and vocational training, as well as work-based learning, can be measured (ETF 2021b). Since 2019, the government has also adopted policies to allow vocational training institutions to offer short-term courses (less than 6 months) to enable continuous vocational education for the workforce. In practice however, there are currently limited opportunities for adult learning due to the lack of adequate adult training facilities and courses, particularly outside of urban centers (ETF 2020b). Azerbaijan could consider online learning platforms to rapidly build up the new skills required by employers.

Under the Republic of Korea's Life-Long Learning Promotion Plan (2018–2022), online learning platforms have been established to upskill the population in a range of areas, including digital skills. A key platform is the Korean Massive Open Online Courses (K-MOOC), which since its launch in 2015, has developed over 1,700 accredited courses at the higher education level through partnerships with local universities, with a significant share of them focused on advanced digital courses such as machine learning, AI navigation and perception, and mathematics for data scientists.[31] In addition, the Ministry of Education in the Republic of Korea runs a Distance University Education program, in which workers have the option to take university courses in new ICT skills, and are awarded degrees upon the completion of such programs.[32]

In Azerbaijan, online learning platforms could similarly be used to upskill the workforce. Nearly all training institutions indicated that they had shifted learning online due to the COVID-19 pandemic, suggesting that many institutions have some experience in this area. The use of online platforms would allow for cheaper course offerings and a wider reach for training courses, circumventing existing issues such as the lack of adult training facilities in rural areas. The Ministry of Education of Azerbaijan could take the lead in developing suitable platforms, working with the State Agency for Public Services and Social Innovations and other relevant agencies.

Recommendation 6: Develop Innovative Job-Matching Initiatives and Platforms

Interviews and consultations with local experts and stakeholders revealed that the lack of information on available jobs and the types of skills required by employers are a key barrier to workers securing quality jobs in Azerbaijan. Students lack sufficient access to information on potential career paths and corresponding skill requirements. Only 4% of training institutions in Azerbaijan work with employers to organize job fairs to advertise job opportunities. Innovative approaches to improve job-matching between employers and prospective workers could address this challenge (see Box 6).

The Ministry of Labor and Social Protection of the Population could consider working with technology agencies such as the State Agency for Public Services and Social Innovations to launch platforms incorporating AI or Big Data technologies to strengthen job matching between workers and firms. Alongside creating job portals, policy makers would also need to embark on outreach efforts and work with industry association, business chambers, as well as training institutions to build awareness of the portals created.

---

[31]   See K-MOOC. Courses. http://www.kmooc.kr/courses.
[32]   See Ministry of Education, Republic of Korea. Lifelong Education. http://english.moe.go.kr/sub/info.do?m=020107&s=english.

Box 6: Developing Artificial Intelligence and Big Data-Powered Job-Matching Platforms in India, Malawi, and Singapore

With the coronavirus disease or COVID-19 pandemic, recruitment fairs have also gone virtual. Platforms such as OnTime Job have built a model where recruiters can finish the whole hiring process on their mobile phone giving them the choice to hire from anywhere and anytime. OnTime Job was launched in India in 2021 and conducted a large-scale virtual recruitment drive with over 8,000 participants and more than 50 successful job matches (*Higher Education Digest* 2020). Using data science and a mobile-first direct hiring approach, Industry 4.0 technologies can save time for recruiters and candidates and will be able to drive better matches compared to traditional hiring processes (*Economic Times* 2020).

Big Data and AI are also used to obtain insights of the labor market in Malawi (ETF 2021a). Over 360,000 data points on myJobo.com, the largest jobsite in Malawi, were analyzed by researchers to provide key insights including the types of trending jobs and the specific skills sought by employers. Furthermore, such analysis supplement existing large-scale surveys like the Malawi Labor Force Survey and provide near-real-time information for jobseekers and training partners to keep up with the labor market trends.

In Singapore, an artificial intelligence (AI)-powered jobs search portal connects jobseekers with career opportunities relevant to their skills. MyCareersFuture.sg seeks to reduce "missed matches" by recommending jobs that are best matched to the job title, skills, and minimum salary desired as indicated by the user. The portal uses machine learning technology to scour various job descriptions posted in the Jobs Bank, and filters the skills required for each type of job. User can then check off the skills they possess and view results of the search, where jobs that best correspond to the skills declared will show up at the top, each labeled with a percentage indicating the extent of match. For instance, a user who searches for the keyword "accountant" will be prompted with a list of relevant skills, such as "financial reporting," "budgets," "account reconciliation," and "Microsoft Excel," among others. The user will also receive recommendations of jobs in other sectors matched to the skills they had provided (*Today* 2018).

Sources: *Today*. 2018. New Jobs Portal Seeks to Reduce Job-Skills Mismatch. 17 April. https://www.todayonline.com/singapore/new-jobs-portal-seeks-reduce-job-skills-mismatch. 2021. European Training Foundation (ETF). 2019. *Big Data for Labour Market Intelligence*. https://www.etf.europa.eu/sites/default/files/2021-06/guide_en_big_data_lmi_etf.pdf; *Higher Education Digest*. 2020. On Time Job Culminates India's Most Successful Live Recruitment Drive. 11 September. https://www.highereducationdigest.com/ontime-job-culminates-indias-most-successful-live-recruitment-drive/. *Economic Times*. 2020. AI-powered Virtual Job Fairs Aim to Create Hassle Free Experience for Job Seekers and Employers. 15 September. https://hr.economictimes.indiatimes.com/news/workplace-4-0/recruitment/ai-powered-virtual-job-fairs-aim-to-create-hassle-free-experience-for-job-seekers-and-employers/78122272.

Recommendation 7: Develop Skilling and Labor Support Programs for Digital Freelancers

One key aspect of the digital economy and 4IR is the proliferation of digital freelancers. Around 15% of young adults in Azerbaijan aged 20–24 years old are unemployed and over 80% of them have at least a secondary education.[33] There is significant potential for young Azerbaijanis to find quality jobs as digital freelancers with the rise of the global freelance economy against the backdrop of the COVID-19 pandemic. A study by Upwork revealed that amidst the pandemic, over a third of the American workforce did freelance work, contributing $1.2 trillion to the United States economy.

[33] State Statistical Committee of Azerbaijan.

In Pakistan, the government has pursued various initiatives to provide skilling opportunities for youth and enable them to work as digital freelancers. The National Freelancing Facilitation Policy 2021 also aims to create a conducive environment for freelance digital work (see Box 7). Similar initiatives could be pursued in Azerbaijan to strengthen the digital freelancing capabilities of its young adults and enable them to find quality freelance jobs. In particular, the Innovation Agency of Azerbaijan (together with the Ministry of Transport, Communication and High Technologies) could lead efforts to provide training to youth in areas such as digital marketing and website development. To create an enabling environment for digital freelancers and remove potential barriers surrounding taxes and fees related to transfer of foreign currencies across borders—as most clients are likely to be based outside of Azerbaijan—the Innovation Agency of Azerbaijan could also consider specific fiscal incentive schemes for freelancers, tapping on Pakistan's example.

Box 7: Various Forms of Government Support for Digital Freelancers in Pakistan

In the Central and West Asia region, Pakistan is home to the world's fourth largest number of digital freelancers working on online platforms for contractual jobs (Geo News 2021). The Pakistan Software Export Board estimates that the exponential growth in the number of freelancers in Pakistan in recent years has resulted in a revenue of $150 million earned by the freelancers in 2019. Government agencies are keen to support such growth, and the Punjab Information Technology Board has launched e-Rozgaar centers to provide training for youth in areas such as e-commerce and website development, which will enable them to work as digital freelancers (see e-Rozgaar homepage).

The Pakistan Ministry of Information Technology and Telecommunication has drafted the National Freelancing Facilitation Policy 2021 to build the number of active freelancers in Pakistan to 1 million (Government of Pakistan, Ministry of Information Technology and Telecommunication 2021). The policy includes programs to develop the skills of local freelancers such as the launch of new training and technology certifications initiatives for freelancers by the Pakistan Software Export Board, fiscal incentives such as income tax holidays, and subsidized health and life insurance for digital freelancers.

Sources: Geo News. 2021. 47% Growth Seen in Pakistan's IT Exports from July–May in Fiscal Year 2020-21. 26 June. https://www.geo.tv/latest/357011-47-growth-seen-in-pakistans-it-exports-from-july-may-in-fiscal-year-2020-21; Startup Pakistan. $150 Million Revenue brought by "Pakistani Freelancers" in a Year. https://startuppakistan.com.pk/150-million-revenue-brought-by-pakistani-freelancers-in-a-year/; e-Rozgaar. About Us. https://www.erozgaar.pitb.gov.pk/#erb04; Government of Pakistan, Ministry of Information Technology and Telecommunication. 2021. *National Freelancing Facilitation Policy 2021 Consultation Draft.* Islamabad. https://moitt.gov.pk/SiteImage/Misc/files/National%20Freelancing%20Facilitation%20Policy%202021%20-%20Consultation%20Draft%202_0.pdf.

Recommendation 8: Develop Programs to Enable More Women to Take Up Technical Jobs

Analysis based on the employer surveys show that while the adoption of 4IR technologies will create net job gains in both the transportation and storage and agro-processing industry, male workers will tend to benefit more from these job gains, which are expected to take place largely in better-paying and more highly skilled technical roles. In contrast, female workers are concentrated in roles that have a higher risk of displacement, such as manual job roles. Past research reveals that while there is upward trend for women pursuing science and technology professions in Azerbaijan, cultural stereotypes continue to pose a barrier for women pursuing technical careers (UNDP 2018). While nearly half of all students enrolled in higher education institutions in Azerbaijan in 2020 were female, only 25% of students enrolled in technical and technological disciplines are female. Instead, most female students are enrolled in education-related disciplines.[34]

---

[34]   State Statistical Committee of Azerbaijan. Education Indicators. https://www.stat.gov.az/source/education/?lang=en.

In Indonesia, the Philippines, and Thailand, the ILO Women in STEM Workforce Readiness and Development Program aims to enhance the employability of women for STEM-related jobs. The program provides training in critical soft and technical STEM-related skills, as well as targeted mentorship opportunities. Participants benefit from tangible employment opportunities upon the completion of the program, which includes activities to identify skills gaps, skill training programs, job placement schemes, and in-company developing and mentoring (ILO 2020a).

In Australia, the Curious Minds program funded by the Department of Education, Skills and Employment is aimed at highly capable female students (aged around 15–16 years old) who have an interest in STEM. This is a 6-month program that combines two residential camps and a coaching program to help ignite girls' passion in STEM. The Government of Australia also funded the development of the Girls in STEM Toolkit that contains resources for female students to find out more about STEM careers, and for teachers to inspire and encourage girls to feel confident and enthusiastic about STEM-related jobs.[35]

Policy makers in Azerbaijan could consider similar initiatives to strengthen the capabilities and interest of women in taking on up technical professions, so that they will benefit from the job gains from 4IR. This could be done in collaboration with international partners such as the ILO. The Center for Analysis and Coordination of the Fourth Industrial Revolution could lead these efforts to ensure that female workers can also benefit from 4IR.

C.   Industry-Specific Priorities

While these recommendations apply to both the agro-processing and transportation and storage industries, a set of priorities unique to each industry should be considered when implementing the respective recommendations.

Agro-Processing Industry

Only 14% of agro-processing firms in Azerbaijan reported a good understanding of 4IR technologies in 2020. Less than 20% of firms expect to adopt technologies such as autonomous robots, additive manufacturing, and Big Data analytics across various functions by 2025 (see Table 9). As such, to reap the gains of 4IR, policy makers would need to play an active role in devising 4IR adoption plans that complement existing sector road maps and support companies to deploy 4IR technologies. The newly established Center for Analysis and Coordination of the Fourth Industrial Revolution could take the lead in coordinating these plans in Azerbaijan, working with key line agencies. An action plan for 4IR technology adoption in the agro-processing industry could be cochaired by the Center and Ministry of Agriculture. The report shows that in making the transition toward 4IR, the lack of quality training providers and impact on women will be particularly challenging for Azerbaijan's agro-processing industry. Targeted programs would need to be adopted to ensure that trainers and teachers in the country are able to support the agro-processing workforce to meet the new skill needs created by 4IR. As roles in the agro-processing industry shift toward more technical occupations dominated by men, active intervention would also be needed to ensure that female workers can benefit from 4IR technologies.

---

[35] Australian Government. *Australian Government Science, Technology, Engineering and Mathematics (STEM) Initiatives for Girls and Women.* https://www.industry.gov.au/sites/default/files/March%202020/document/australian-government-stem-initiatives-for-women-and-girls-2019.pdf.

Table 9: Summary of Findings in the Agro-Processing Industry

Key Findings

Potential job displacement (% of current workforce): 30,200 (42%)
Potential job gains (% of current workforce): 45,000 (62%)
Net job gains from 4IR (% of current workforce): 14,800 (20%)
Top three in-demand skills in 2025: Digital and/or information and communication technology skills, creative thinking and/or design, critical thinking

Key Challenges	Findings	Recommendations
Limited understanding of 4IR technologies and their applications	Apart from IOT technologies, adoption of 4IR technologies is expected to remain low among agro-processing firms in 2025	Develop 4IR adoption plans to complement existing sectoral road maps
Lack of good quality training providers	Over **60%** of firms disagree that it is easy to find good quality trainers.	Programs to strengthen industry knowledge and digital skills of trainers and teachers
Job gains from 4IR benefit male workers more than female workers	The number of new jobs expected to be gained by male workers is **1.6 times** that expected to be gained by female workers	Develop programs to enable more women to take up technical jobs

4IR = Fourth Industrial Revolution, IOT = Internet of Things.
Source: Asian Development Bank (Sustainable Development and Climate Change Department).

Transportation and Storage Industry

Like the agro-processing industry, transportation and storage companies in Azerbaijan also have a limited understanding of 4IR, with only 20% of firms surveyed reporting a good understanding of 4IR and their applications (see Table 10). However, a larger proportion of firms expect to adopt various 4IR technologies such as systems integration, blockchain technology, and IOT technology by 2025. As such, it would be integral for policy makers to support firms in making the transition toward 4IR by developing programs and 4IR competency centers to build awareness of digital tools. In tandem with the shift toward 4IR technology adoption, the job scope of workers in the transportation and storage industry will change as would skill needs. Policy makers could support the reskilling of workers through incentive schemes for firms to train employees and the development of online learning platforms. In particular, grants could be provided to companies to cover training fees of workers, and absentee payroll funding could be offered to cover a proportion of the worker's salary during the training period to ensure that employers are not deterred by the loss of human resources when employees undergo training. Programs modeled after Singapore's job redesign reskilling and/or redeployment programs could support the transition of firms toward 4IR. These programs support firms that are undergoing business transformation, and have existing workers at risk of redundancy or in vulnerable jobs due to the transformation. Support is provided to allow workers to be trained to take on new job roles or redesigned job roles within the company.[36]

---

[36]   Skills Future Singapore. Job Redesign Reskilling/Redeployment Programs. https://www.wsg.gov.sg/programmes-and-initiatives/job-redesign-reskilling-redeployment.html.

Table 10: Summary of Findings in the Transportation and Storage Industry

Key Findings

Potential job displacement (% of current workforce): 66,500 (32%)
Potential job gains (% of current workforce): 93,800 (46%)
Net job gains from 4IR (% of current workforce): 27,300 (13%)
Top three in-demand skills in 2025: Digital and/or information and communication technology skills, creative thinking and/or design, numeracy

Key Challenges	Findings	Recommendations
There is limited understanding of the applications of 4IR technologies among firms	Only **20%** of transportation and storage firms have a good understanding of 4IR technologies and applications	Develop programs and 4IR competency centers to build awareness of digital tools among firms
Most jobs created will be in technical roles that the workforce might not be equipped for	**32%** of transportation and storage firms expect the number of technical jobs to increase	Implement incentive schemes for firms to train employees for 4IR
Significant changes in skill demand may lead to challenges in hiring workers	Digital and/or information and communication technology skills, creative thinking and/or design, and numeracy skills will become more important with 4IR adoption	Develop online learning platforms

4IR = Fourth Industrial Revolution.
Source: Asian Development Bank (Sustainable Development and Climate Change Department).

APPENDIX
Participants Engaged During National Consultations

The following stakeholders were engaged in initial consultations for Azerbaijan:

	Government Ministries and Agencies		
1	Tamerlan Tagiyev	Acting Executive Director	Center for Analysis and Coordination of the Fourth Industrial Revolution, Ministry of Economy
2	Gulmina Malikzade	Representative	
3	Huseyn Guliyev	Head, International Donor Cooperation Sector, Department of Cooperation with International Organizations	Ministry of Economy
4	Najiba Tagiyeva	Specialist, Industrial Policy Sector of the Industry Department	
5	Tural Valiyev	Head, Strategy and Corporate Development Department	Secretariat of the Small and Medium Business Development Agency, Ministry of Economy
6	Narmin Shahbazova	Chief Advisor	Azerbaijan State Vocational Education Agency, Ministry of Education
7	Heydar Agayev	Chief Specialist, Department of Youth Affairs	Ministry of Youth and Sport
8	Semral Aliyev	Head of Sector, Innovation Sector, Department of e-Government and Digital Innovations	State Agency for Public Services and Social Innovations
9	Javid Abbasov	Lead Consultant, E-Government Sector, E-Government and Digital Innovations Department	
10	Inji Jafarli	Chief Specialist, International Relations Department	
11	Laman Salimova	International Cooperation Specialist of the Innovation Center	
12	Rashad Huseynov	Head, Economic Analysis and Analytical Information Department	Centre for Analysis of Economic Reforms and Communication

continued on next page

Table *continued*

13	Leyla Mammadova	Deputy Chairman, Agrarian Credit and Development Agency	Ministry of Agriculture
14	Farid Amirov	Head of the Projects and Programs Department, Agrarian Credit and Development Agency	
15	Gunel Jumshudova	Chief Specialist of the Farmers' Information and Market Information Division of the Analytical Department, Agrarian Research Center	
16	Samir Rustamli	Head of Strategic Analysis and Planning Sector	Azerbaijan Food Safety Agency
Nongovernment Stakeholders			
17	Ehtiram Muslumov	Senior Specialist, International Relations and Sustainable Development Department	National Confederation of Entrepreneurs (Employers) Organizations
18	Nurlana Hasanova	Senior Specialist, Entrepreneurs and Communication Department	
19	Ganira Moylamova	International Relations Specialist	
20	Nurlana Mahmudova	Head of the Department for Work with Students	Baku State Vocational Education Center on Industry and Innovations
21	Orkhan Rahimov	Deputy Director	
22	Ruslan Atakishiyev	Director of Innovative Business Incubator LLC	Azerbaijan State University of Economics
23	Shabnam Safarova	Senior Specialist of International Relations Department	Baku State University

The following stakeholders were engaged in further consultations for Azerbaijan:

1	Ministry of Economy	Huseyn Guliyev	Head, International Donors Cooperation Sector of the Cooperation with International Organizations Department
		Najiba Tagiyeva	Senior consultant, industrial policy, Department of Industry
2	Small and Medium Business Development Agency	Emil Mammadov	Head, SME Market Support Department
3	Center for Analysis and Coordination of the Fourth Industrial Revolution	Tamerlan Tagiyev	Head, Center for Coordination and Analysis of the Fourth Industrial Revolution
4	Ministry of Labor and Social Protection of Population	Shalala Mardaliyeva	Senior Adviser, Employment Policy and Demographic Development Department
5	Centre for Analysis of Economic Reforms and Communication	Rashad Huseynov	Head, Economic Analysis and Analytical Information Department
6	Azerbaijan Food Safety Agency	Samir Rustamli	Head, Strategic Analysis and Planning Sector

continued on next page

Table *continued*

7	Ministry of Agriculture	Farid Amirov	Head of the Projects and Programs Department, Agrarian Credit and Development Agency
		Hasanov Javid	Chief Specialist, Agrarian Market Development and Agribusiness Environment Improvement Department of the Research Department, Agrarian Research Center
		Mahir Hatamov	Leading Specialist, Diagnostics, Forecasting and Planning Division of the Analytical Department, Center for Agrarian Research
8	Ministry of Transport, Communications and High Technologies	Ibrahim Aliyev	Deputy Head, Transport Policy Department
		Alish Ismayilov	Leading Consultant, Transport Policy Department
		Shamsiyya Guliyeva	Consultant, Transport Policy Department
		Asif Hasanov	Senior Advisor, Transport Regulation Department
9	Ministry of Education	Nijat Mammadli	Head, Department of Science, Higher and Secondary Special Education
Industry Associations			
1	National Confederation of Entrepreneurs (Employers) Organizations	Kristina Mammadova	Acting hief Executive Officer
		Fuad Humbatov	Deputy Secretary General
		Ehtiram Muslimov	Senior Specialist, International Relations and Sustainable Development
		Nurlan Musayev	Senior Specialist, International Relations and Sustainable Development
		Ganira Moylamova	Senior Specialist, International Relations and Sustainable Development
		Shebnem Nuriyeva	Adviser to the President, Director of ASK Business Consulting
		Aysel Yusifova	Member of ASK Commission on Management and Business Transformation
2	German Azerbaijan Chamber of Commerce	Magnus Muller	Head of Vocational Education

References

Asian Development Bank. 2021. *Reaping the Benefits of Industry 4.0 through Skills Development in Viet Nam.* Manila. https://www.adb.org/publications/benefits-industry-skills-development-viet-nam.

———. 2020. *Azerbaijan Moving Toward More Diversified, Resilient, and Inclusive Development.* Manila. https://www.adb.org/publications/azerbaijan-diversified-resilient-inclusive-development.

AIG. 2017. *IOT Case Studies: Companies Leading the Connected Economy.* https://www.indevagroup.com/wp-content/uploads/2017/12/iot-case-studies-companies-leading-the-connected-economy-digital-report.pdf.

Aliyev, S. 2019. Problems and Opportunities for Leveraging SME Finance through Value Chains in Azerbaijan. *ADBI Working Paper Series.* No. 973. Tokyo: Asian Development Bank Institute. https://www.adb.org/sites/default/files/publication/511266/adbi-wp973.pdf.

AlphaBeta. 2017. *The Automation Advantage.* https://alphabeta.com/wp-content/uploads/2017/08/The-Automation-Advantage.pdf.

AMFG. How 3D Printing Transforms the Food and Beverage Industry. https://amfg.ai/2020/08/17/how-3d-printing-transforms-the-food-and-beverage-industry/.

Azerbaijan Press Agency. 2021. Center for Analysis and Coordination of the Fourth Industrial Revolution under the Ministry of Economy Established. 6 January. https://apa.az/en/infrastructure/Center-for-Analysis-and-Coordination-of-the-Fourth-Industrial-Revolution-under-the-Ministry-of-Economy-established-339541.

AzerNews. 2020. Azerbaijani Agency for Development of SMEs Informs Entrepreneurs about E-commerce Development. 15 May. https://www.azernews.az/business/165125.html.

———. 2021a. Azerbaijan to Launch ICT Program Based on International Experience. 13 January. https://www.azernews.az/nation/174966.html.

Azertag. 2021. *Order of the President of the Republic of Azerbaijan on Approval of "Azerbaijan 2030: National Priorities for Socio-Economic Development.* 2 February. https://azertag.az/en/xeber/Order_of_the_President_of_the_Republic_of_Azerbaijan_on_approval_of_Azerbaijan_2030_National_Priorities_for_Socio_Economic_Development-1724707.

Banker, S. 2020. Robots and The Autonomous Supply Chain. *Forbes.* 2 April. https://www.forbes.com/sites/stevebanker/2020/04/02/robots-and-the-autonomous-supply-chain/?sh=246eafea787a.

Barber. 2007. *Instruction to Deliver: Fighting to Transform Britain's Public Services.* https://books.google.com.sg/ books/about/Instruction_to_Deliver.html?id=MLcbAQAAMAAJ&redir_esc=y

BDO Tax News. Azerbaijan Recent Tax Changes. https://www.bdo.global/en-gb/microsites/tax-newsletters/ corporate-tax-news/issue-50-february-2019/azerbaijan-recent-tax-changes.

BMWI. 2019. *Case Study on the Mittelstand 4.0 Competence Centres, Germany: Contribution to the OECD TIP Digital and Open Innovation Project.* Berlin. https://www.innovationpolicyplatform.org/www. innovationpolicyplatform.org/system/files/imce/SME4.0CompetenceCentres_Germany_ TIPDigitalCaseStudy2019_1/index.pdf.

Center for Economic and Social Development. 2020. *Amendments to State Budget of Azerbaijan for 2020: Reasons and Expectations.* https://cesd.az/new/wp-content/uploads/2020/08/Amendments_State_Budget_2020_ Azerbaijan.pdf.

Channel News Asia. 2021. Redefine Meat Raises US$29 Million to Finance Rollout of 3D-Printed Meat Substitute. 16 February. https://www.channelnewsasia.com/news/business/redefine-meat-raises-us-29-million-to-finance-rollout-of-3d-printed-meat-substitute-14208178.

Department of Environmental Health. 2001. *Musculoskeletal Risks in Washington State Apple Packing Companies.* Seattle. https://deohs.washington.edu/sites/default/files/images/general/applepacking.pdf.

DHL. 2018. *Blockchain in Logistics.* https://www.dhl.com/content/dam/dhl/global/core/documents/pdf/glo-core-blockchain-trend-report.pdf.

———. Artificial Intelligence: AI. Today a Novelty, Tomorrow a Necessity. https://www.dhl.com/sg-en/home/ insights-and-innovation/insights/artificial-intelligence.html.

———. Digital Twins. https://www.dhl.com/global-en/home/insights-and-innovation/thought-leadership/trend-reports/virtual-reality-digital-twins.html

———. Internet of Things. https://www.dhl.com/global-en/home/insights-and-innovation/thought-leadership/ trend-reports/internet-of-things-in-logistics.html.

e-Rozgaar. About Us. https://www.erozgaar.pitb.gov.pk/#erb04.

Economic Times. 2020. AI-powered Virtual Job Fairs Aim to Create Hassle Free Experience for Job Seekers and Employers. 15 September. https://hr.economictimes.indiatimes.com/news/workplace-4-0/ recruitment/ai-powered-virtual-job-fairs-aim-to-create-hassle-free-experience-for-job-seekers-and-employers/78122272.

Education and Training Foundation. Taking Teaching Further. https://www.et-foundation.co.uk/supporting/ support-for-teacher-recruitment/taking-teaching-further/.

Enterprise Singapore. Productivity Solutions Grant. https://www.enterprisesg.gov.sg/financial-assistance/grants/ for-local-companies/productivity-solutions-grant.

European Training Foundation (ETF). 2014. *Sector Skills Councils in Azerbaijan.* Turin. https://www.etf.europa.eu/ sites/default/files/2018-10/Sector%20skills%20councils%20in%20Azerbaijan.pdf.

———. 2018. *Azerbaijan Country Strategy Paper 2017-20: 2018 Updates*. Turin. https://www.etf.europa.eu/sites/default/files/m/96B9F9A8EA4FF770C125821F005391A7_CSP%202017-2020%20AZERBAIJAN_Updates%202018.pdf.

———. 2020a. *Torino Process 2018–2020 Azerbaijan National Report*. Turin. https://openspace.etf.europa.eu/sites/default/files/2020-09/TPRreport%202019%20Azerbaijan_EN.pdf.

———. 2020b. *Policies for Human Capital Development Azerbaijan. Turin*. https://www.etf.europa.eu/en/publications-and-resources/publications/trp-assessment-reports/azerbaijan-2020.

———. 2021a. *Big Data for Labour Market Intelligence*. https://www.etf.europa.eu/sites/default/files/2021-06/guide_en_big_data_lmi_etf.pdf.

———. 2021b. *National Qualifications Framework – Azerbaijan*. https://www.etf.europa.eu/sites/default/files/document/Azerbaijan.pdf.

Fau, S. and Moreau, Y. 2018. *Managing Tomorrow's Digital Skills: What Conclusions can We Draw from International Comparative Indicators?* United Nations Educational, Scientific and Cultural Organization (UNESCO) Digital Library. https://unesdoc.unesco.org/ark:/48223/pf0000261853.

Geo News. 2021. 47% Growth Seen in Pakistan's IT exports from July–May in Fiscal Year 2020-21. 26 June. https://www.geo.tv/latest/357011-47-growth-seen-in-pakistans-it-exports-from-july-may-in-fiscal-year-2020-21.

German-Azerbaijani Chamber of Commerce (AHK Azerbaijan) and KPMG Azerbaijan. 2020. *Foreign Business in Azerbaijan Report 2020*. https://www.aserbaidschan.ahk.de/en/marktinformation/publikationen/ahk-aserbaidschan-publikationen/auslandsgeschaeft-in-aserbaidschan-berichte/foreign-business-in-azerbaijan-report-2020.

GIZ. Private Sector Development and Technical Vocational Education and Training, South Caucasus. https://www.giz.de/en/worldwide/20324.html.

Government of Australia. *Australian Government Science, Technology, Engineering and Mathematics (STEM) Initiatives for Girls and Women*. https://www.industry.gov.au/sites/default/files/March%202020/document/australian-government-stem-initiatives-for-women-and-girls-2019.pdf.

Government of the Republic of Azerbaijan. Ministry of Transport, Communication and High Technologies. First Internet of Things Laboratory Created in Azerbaijan. https://mincom.gov.az/en/view/news/590/first-internet-of-things-laboratory-created-in-azerbaijan; and country consultation workshop.

———. 2016a. *Strategic Vision and Roadmap for Azerbaijan Agriculture*. Baku. https://monitoring.az/assets/upload/files/8047fecde10eaf0fd8cb45de716d8267.pdf.

———. 2016b. *Strategic Roadmap for the Development of Logistics and Trade in Azerbaijan*. Baku. https://monitoring.az/assets/upload/files/4eae769862be45d63dcd5b50b1d31844.pdf

———. 2016c. *Strategic Roadmap for Vocational Education and Training in Azerbaijan Republic*. Baku. https://monitoring.az/assets/upload/files/6381dda5389fb17755bbb732a9c7d708.pdf.

———. 2016d. *Strategic Roadmap for National Economy Perspective of the Republic of Azerbaijan.* Baku. https://monitoring.az/assets/upload/files/15075d6928310402cd152c96db0d6835.pdf.

———. 2016e. *Strategic Roadmap for Development of Telecommunications and Information Technologies in Azerbaijan Republic.* Baku. https://monitoring.az/assets/upload/files/6683729684f8895c1668803607932190.pdf.

———. 2016f. *Presidential Decree of the Azerbaijan Republic about Approval of Strategic Road Maps on National Economy and the Main Sectors of Economy.* Baku. https://cis-legislation.com/document.fwx?rgn=91715.

———. 2019. *Employment Strategy 2019–2030 (2019–2030-cu illər üçün Azərbaycan Respublikasının Məşğulluq Strategiyası).* Baku. http://e-qanun.az/framework/40416.

Government of Pakistan, Ministry of Information Technology and Telecommunication. 2021. *National Freelancing Facilitation Policy 2021 Consultation* Draft. Islamabad https://moitt.gov.pk/SiteImage/Misc/files/National%20Freelancing%20Facilitation%20Policy%202021%20-%20Consultation%20Draft%202_0.pdf.

Government of the Republic of Korea, Ministry of Education. Lifelong Education (in Korean). http://english.moe.go.kr/sub/info.do?m=020107&s=english.

Government of Singapore, Ministry of Trade and Industry. ITMs Logistics. https://www.mti.gov.sg/ITMs/Trade_Connectivity/Logistics.

———. ITMs Overview. https://www.mti.gov.sg/ITMs/Overview.

Groenfeldt, T. 2017. IBM And Maersk Apply Blockchain to Container Shipping. *Forbes.* 5 March. https://www.forbes.com/sites/tomgroenfeldt/2017/03/05/ibm-and-maersk-apply-blockchain-to-container-shipping/?sh=febf8443f05e.

Higher Education Digest. 2020. On Time Job Culminates India's Most Successful Live Recruitment Drive. 11 September. https://www.highereducationdigest.com/ontime-job-culminates-indias-most-successful-live-recruitment-drive/.

IBM. 2018. Maersk and IBM Introduce TradeLens Blockchain Shipping Solution. https://newsroom.ibm.com/2018-08-09-Maersk-and-IBM-Introduce-TradeLens-Blockchain-Shipping-Solution.

———. What is Blockchain Technology? https://www.ibm.com/sg-en/topics/what-is-blockchain.

IEEE. 2021. *Secure Transport of COVID-19 Vaccines with IOT Cold Chain Monitoring.* 20 January https://innovate.ieee.org/innovation-spotlight/cold-chain-monitoring/.

IMDA. Industry Digital Plans. https://www.imda.gov.sg/programme-listing/smes-go-digital/industry-digital-plans.

———. *Logistics Industry Digital Plans.* https://www.imda.gov.sg/programme-listing/smes-go-digital/industry-digital-plans/logistics-idp#:~:text=Aligned%20to%20the%20Logistics%20Industry,each%20stage%20of%20their%20growth.

International Labour Organization (ILO). 2020a. Women in STEM Workforce Readiness and Development Programme. Geneva. https://www.ilo.org/asia/projects/WCMS_619723/lang--en/index.htm.

———. 2020b. *ILO Toolkit for Quality Apprenticeships*. Geneva. https://www.ilo.org/wcmsp5/groups/public/---ed_emp/---ifp_skills/documents/publication/wcms_751114.pdf.

INTI International University & Colleges. Industry Attachment Programme Redefines Teaching at INTI. https://newinti.edu.my/industry-attachment-programme-redefines-teaching-at-inti/.

K-MOOC. Courses. http://www.kmooc.kr/courses.

Kim et al. 2017. Korea's Software Education Initiative. https://dl.acm.org/doi/abs/10.1145/3151759.3151800?download=true.

Maddox, T. 2017. How Hershey used IOT to Save $500K for Every 1% of Improved Efficiency in Making Twizzlers. *Tech Republic*. 24 February. https://www.techrepublic.com/article/how-hershey-used-the-cloud-to-deploy-iot-and-machine-learning-without-a-data-scientist/.

Marr, B. 2017. The Amazing Ways Coca Cola Uses Artificial Intelligence and Big Data to Drive Success. *Forbes*. 18 September. https://www.forbes.com/sites/bernardmarr/2017/09/18/the-amazing-ways-coca-cola-uses-artificial-intelligence-ai-and-big-data-to-drive-success/?sh=5aba7a6378d2.

———. 2018. The Brilliant Ways UPS Uses Artificial Intelligence, Machine Learning and Big Data. *Forbes*. 15 June. https://www.forbes.com/sites/bernardmarr/2018/06/15/the-brilliant-ways-ups-uses-artificial-intelligence-machine-learning-and-big-data/?sh=3c3398215e6d.

McKinsey & Company. 2012. Delivery 2.0: The New Challenge for Governments. https://www.mckinsey.com/industries/public-sector/our-insights/delivery-20-the-new-challenge-for-governments.

———. 2016. Supply Chain 4.0—The Next-Generation Digital Supply Shain. https://www.mckinsey.com/business-functions/operations/our-insights/supply-chain-40--the-next-generation-digital-supply-chain.

———. 2018. *Food Processing & Handling Ripe for Disruption?* https://www.mckinsey.com/~/media/mckinsey/industries/advanced%20electronics/our%20insights/whats%20ahead%20for%20food%20processing%20and%20handling/mckinsey-on-food-processing-and-handling-ripe-for-disruption.ashx.

———. 2020. *How COVID-19 has Pushed Companies over the Technology Tipping Point—and Transformed Business Forever.* https://www.mckinsey.com/business-functions/strategy-and-corporate-finance/our-insights/how-covid-19-has-pushed-companies-over-the-technology-tipping-point-and-transformed-business-forever.

McKinsey Global Institute. 2014. *Southeast Asia at the Crossroads: Three Paths to Prosperity.* https://www.mckinsey.com/~/media/McKinsey/Featured%20Insights/Asia%20Pacific/Three%20paths%20to%20sustained%20economic%20growth%20in%20Southeast%20Asia/MGI%20SE%20Asia_Executive%20summary_November%202014.ashx.

Microsoft and AlphaBeta. 2019. *Preparing for AI: The implications of Artificial Intelligence for Jobs and Skills in Asian Economies.* https://news.microsoft.com/apac/2019/08/26/preparing-for-ai-the-implications-of-artificial-intelligence-for-jobs-and-skills-in-asian-economies/.

MIT Technology Review. 2018. How UPS Uses AI to Deliver Holiday Gifts in the Worst Storms. 21 November. https://www.technologyreview.com/2018/11/21/139000/how-ups-uses-ai-to-outsmart-bad-weather/.

National Center on Education and the Economy. 2020. *South Korea: Learning Systems*. http://ncee.org/what-we-do/center-on-international-education-benchmarking/top-performing-countries/south-korea-overview/south-korea-instructional-systems/.

Ng, D. and Enriquez, M. 2020. How 3D Food Printing Can Help the Elderly. *Channel News Asia*. 26 February. https://www.channelnewsasia.com/news/cnainsider/how-3d-food-printing-can-help-the-elderly-nutrition-12470760.

Organisation for Economic Co-operation and Development (OECD). 2018. *Getting Skills Right: Future-Ready Adult Learning Systems.* Paris. https://read.oecd-ilibrary.org/education/getting-skills-right-future-ready-adult-learning-systems_9789264311756-en#page1.

PISA. 2012. Main Results from the PISA 2012 Computer-Based Assessments. https://www.oecd-ilibrary.org/docserver/9789264239555-6-en.pdf?expires=1649834501&id=id&accname=guest&checksum=45BD7094B35E5CA72103B75AF6B7D1E3.

Port of Baku. 2021. In 2020, the Cargo Turnover of Baku Port will Increase by 20 Percent. 18 January. http://portofbaku.com/MediaCenter/News/1083.

PWC. 2018. *The Macroeconomic Impact of Artificial Intelligence*. https://www.pwc.co.uk/economic-services/assets/macroeconomic-impact-of-ai-technical-report-feb-18.pdf.

Sharma, S. 2019. How Artificial Intelligence is Revolutionizing Food Processing Business? *Towards Data Science*. 15 July. https://towardsdatascience.com/how-artificial-intelligence-is-revolutionizing-food-processing-business-d2a6440c0360.

Ship Technology. 2017. Could Blockchain Technology Revolutionise Shipping? 10 September. https://www.ship-technology.com/features/featurecould-blockchain-technology-revolutionise-shipping-5920391/.

Singapore Institute of Purchasing and Materials Management. 2019. *Six Essential Areas for an Integrated Logistics System.* https://publication.sipmm.edu.sg/six-essential-areas-integrated-logistics-system/.

Skills Future Singapore. Job Redesign Initiatives. https://www.ssg-wsg.gov.sg/employers/job-redesign.html.

SME Corp Malaysia. SME's Skills Upgrading Programme. https://www.fmm.org.my/images/articles/branches/Negeri_Sembilan/SME%20PROG.pdf.

South East Farmer. 2021. Robotic Packer at the 'Core' of Apple Producer's Automation. 3 February. https://www.southeastfarmer.net/section/fruit/robotic-packer-at-the-core-of-apple-producers-automation.

Startup Pakistan. Motive's AI Conference. https://startuppakistan.com.pk/150-million-revenue-brought-by-pakistani-freelancers-in-a-year/.

State Statistical Committee of Azerbaijan. Education Indicators. https://www.stat.gov.az/source/education/?lang=en.

———. Industry of Azerbaijan. https://www.stat.gov.az/source/industry/?lang=en.

———. Labour Market. https://www.stat.gov.az/source/labour/?lang=en.

———. System of National Accounts and Balance of Payments. https://www.stat.gov.az/source/system_nat_ accounts/?lang=en.

Telecoms. 2015. Ericsson and Inmarsat to Help Global Fleets to Run a Tighter Ship. 19 November. https://telecoms.com/455062/ericsson-and-inmarsat-to-help-global-fleets-to-run-a-tighter-ship/.

The Kewpie Group. 2020. *Integrated Report 2020.* https://www.kewpie.com/en/ir/pdf/integrated-report/ Integrated_Report_2020.pdf.

Today. 2018. New Jobs Portal Seeks to Reduce Job-Skills Mismatch. 17 April. https://www.todayonline.com/ singapore/new-jobs-portal-seeks-reduce-job-skills-mismatch.

United Nations Development Programme (UNDP). 2018. W*omen in the Private Sector in Azerbaijan: Opportunities and Challenges. New York City.* https://azerbaijan.unfpa.org/sites/default/files/pub-pdf/UNDP-AZE-Gender-Assessmnet-kitab-eng_v4_view.pdf.

UNDP Azerbaijan. 2020a. Ministry of Labor and Social Protection of the Population, UNDP and Coursera Join Forces to Support 50000 Azerbaijani Citizens Who Lost Their Jobs during COVID-19 Gain New Skills and Knowledge. 30 September. https://www.az.undp.org/content/azerbaijan/en/home/presscenter/news/2020/ UNDP-AZE-Coursera-Workforce-Recovery-Initiative.html.

———. 2020b. A One-Stop Digital Platform Making E-Services Available to Everyone During Coronavirus Pandemic. 6 April. https://www.az.undp.org/content/azerbaijan/en/home/presscenter/pressreleases/2018/ UNDP-AZE-evdeqalaz-launched.html.

———. 2021. Women and Girls in STEM: New Mentorship Programme Launched in Azerbaijan. 25 January. https://www.undp.org/azerbaijan/press-releases/women-and-girls-stem-new-mentorship-programme-launched-azerbaijan.

University of Nottingham. 2016. *Artificially-intelligent Cleaning System Could Save Food Manufacturers £100m a Year.* https://www.nottingham.ac.uk/news/pressreleases/2016/september/new-ai-driven-cleaning-system-could-save-food-manufacturers-100m-a-year.aspx.

Versai, A. 2021. Walmart Warehouses to be Staffed by Autonomous Robots. *Technowize.* 29 January. https://www.technowize.com/walmart-warehouses-to-be-staffed-by-autonomous-robots/.

World Bank. Unemployment Data. https://data.worldbank.org/indicator/SL.UEM.TOTL.ZS?locations=AZ (accessed 15 September 2021).